THE GOLDEN LEAF

The Golden Leaf

How Tobacco Shaped Cuba and the Atlantic World

CHARLOTTE A. COSNER

VANDERBILT UNIVERSITY PRESS
Nashville

Nashville, Tennessee 37235

First printing 2014

This book is printed on acid-free paper.
Text design and composition by Tseng Information Systems, Inc.
Manufactured in the United States of America

Library of Congress Cataloging-in-Publication Data on file

LC control number 2014013076
LC classification HD9144.C92C67 2014
Dewey class number 338.4′76797097291—dc23

ISBN 978-0-8265-2032-6 (hardcover)
ISBN 978-0-8265-2034-0 (ebook)

In memory of Raymond L. Cosner and Lester V. Cosner,
the two men who taught me about tobacco.

Contents

Acknowledgments

OVER THE COURSE OF THIS PROJECT, I have amassed a list of people who have aided me in innumerable ways, and there are many more I have not mentioned by name here.

Funding from various sources made research trips to distant archives and conferences possible. I gratefully acknowledge the financial support for this project from the following institutions: the Program for Cultural Cooperation between Spain's Ministry of Culture and US Universities, the Lydia Cabrera Award for Cuban Historical Studies from the Conference on Latin American History, the Ford Foundation Grant for Student Travel from the Cuban Research Institute at Florida International University, the Escuela de Estudios HispanoAmericanos, the Jay I. Kislak Foundation, Florida International University's Department of History, and the Chancellor's Office and the History Department at Western Carolina University.

Staff and directors at the Archivo General de Indias and the Centro de Estudios HispanoAmericanos in Seville, Spain, the Archivo Nacional de Cuba, the Archivo Historico Provincial de Pinar del Río, the Special Collections Department at the Steven and Dorothea Green Library at Florida International University, the Special Collections Department at the Otto G. Richter Library at the University of Miami, and the Interlibrary Loan staff at Western Carolina University all provided a wonderful atmosphere for research, granted access to the materials needed to bring this work to completion, and were ever patient with my many requests.

My infinite gratitude and thanks go to my dissertation advisor and mentor, Sherry Johnson at Florida International University. Sherry helped me with this topic long before I officially became her student, and continued to do so even after I had stopped. Along the way, she taught me to be a more efficient and dedicated scholar and a cleaner writer, and she became a lifelong friend. I'm lucky to have Sherry in my corner and hope that this long-overdue work can serve as partial compensation for all that she has done on my behalf over the years.

At archives and conferences, I have discussed various aspects of this topic with numerous people, all of whom have given me great insights and asked probing questions. I cannot name them all, but am very grateful for their time and feedback. I owe a debt to the anonymous readers who made invaluable suggestions and comments. The editorial board and staff of Vanderbilt University Press have shepherded me through the publication process with such kindness, patience, and helpfulness and have made every step an absolute joy. Eli Bortz, in particular, expressed a very early interest in this work, and showed the patience of Job as I produced this final version.

Friends and family tolerated my increasing absences, both physically and mentally, as I plodded away at the book, writing and rewriting it more times than I can count. They encouraged me, silently at times, and more vocally toward the end of the project when I needed it most. In Cullowhee, I'm proud to call a group of pretty amazing people my friends. They have made the task of focusing on my job of "Working Mom" that much easier. During the summer of 2012, in particular, Nora Leal, Guillermo Cardona, Matilde "Mati" Castellanos, and Viviana Alferez provided me with the luxury of living as a full-time writer, which meant I was able to advance this work in ways that would otherwise have been impossible. My children have grown up with this project. Patrick, Virginia, and Adam have never known a time, it seems, when Mommy wasn't working on her book. Maybe when they are finally able to hold it in their hands the idea will become real, and they will overlook the many times when I went to the office and stayed long past their bedtimes. I am so grateful for their patience with me and their understanding, as well as that of my husband, Rick Cardona, who has remained my rock through thick and thin.

Introduction

CHRISTOPHER COLUMBUS MADE CLEAR and careful observations of the local geography, its inhabitants, and their customs during his 1492 voyage to the New World. In early November, Columbus sent two men, Rodrigo de Jerez and Luis de Torres, to explore the interior of an island. The crewmembers returned within a week, bringing tales of a weed twisted and smoked by the Native Americans. Europeans had seen this dried substance earlier, during encounters with the area's indigenous populations. Although Columbus and his crew were uncertain of the plant's name, these descriptions provide the first known European reference to tobacco. The significance of Europe's introduction to this important commodity is clear today, but at the time, Columbus and his men were unimpressed by this new discovery.[1]

Gold, in contrast, was of great interest. Columbus's journal noted the items traded by the indigenous people, such as "parrots and cotton thread in balls and javelins and many other things," but what the sailor hoped to find was gold. The Spanish sovereigns Ferdinand and Isabel had granted the Genoese sailor many rights and privileges under the "Capitulations of Santa Fé." These included the ranks of admiral, viceroy, and governor-general, along with 10 percent profit on all goods obtained by trade. Columbus therefore was under a tremendous amount of both personal and external pressure to succeed in his overseas mission. Columbus's own log detailed this struggle. Columbus anchored on 15 October off the coast of an island he named Santa María de la

Concepción, and searched for gold. Finding little of value, he headed to another island "where all these men that I am bringing from San Salvador make signs that there is very much gold and that they wear rings of it on their arms and on their legs and in their ears and on their noses and on their chests." Columbus was particularly hopeful that gold would be found because the Native Americans' attire indicated that the metal was available and, most likely, nearby. Day after day, his journal entries note his continued pursuit of the precious metal. Although he and his men searched diligently, they failed to find the gold source they were expecting, despite consistent Native American reports that there was gold on their own island or not far away. Columbus's focus remained singular, and he admitted that "there may be many things that I do not know about because I do not want to stop, so I can investigate and go to many islands in order to find gold." It became clear to the Europeans by 2 January 1493, that they would find no substantial sources of the desirable metal on this trip. Columbus refused to give up. He left thirty-nine men on the island of Hispaniola, hoping that they would find the elusive source of the little gold observed during the voyage.[2]

Neither Cuba nor the other Caribbean islands possessed the quantities of gold that European conquistadors hoped to discover. Still, the quest for gold persisted among New World sixteenth-century explorers and adventurers as Spain continued to search for the precious metal on subsequent expeditions to the Mesoamerican and South American mainland. Some were successful, like Hernán Cortés and Francisco Pizarro in Mexico and Peru, respectively. Others—such as Hernando de Soto and Francisco Vázquez de Coronado, at opposite ends of what is now the southern part of the United States—were not. In the wake of the discovery of the Aztec and Inca civilizations' vast treasure, the Spaniards' drive was understandable. Initially, Europeans paid little attention to other economic activities, overlooking other possibilities that could have proven more lucrative, Georg Friederici argues. The search for and obtaining of mineral resources became a standard part of Iberian conquest and colonization. Gold and other precious minerals obtained through Native American labor were a seemingly endless resource and provided Spain with riches. Unfortunately for Spain, however, the cost of empire was great, and expenses frequently outpaced the New World's income, causing Spain's Felipe II (reigned 1556–1598) to declare bankruptcy no less than four times within a span of less than fifty years.[3]

Columbus may have returned home without the glittering treasure he had hoped to find, yet his first voyage was far from unproductive. Exploration led to expanded European empires, and over time, tobacco became an important New World cash crop. From a botanical perspective, however, tobacco seems rather inauspicious. The plant is a member of the large *Solanaceæ* family, which includes 96 genera and 2,297 species, such as nightshade, Irish potato, jimsonweed, chili pepper, tomato, petunia, and eggplant. Each of the 95 known tobacco varieties

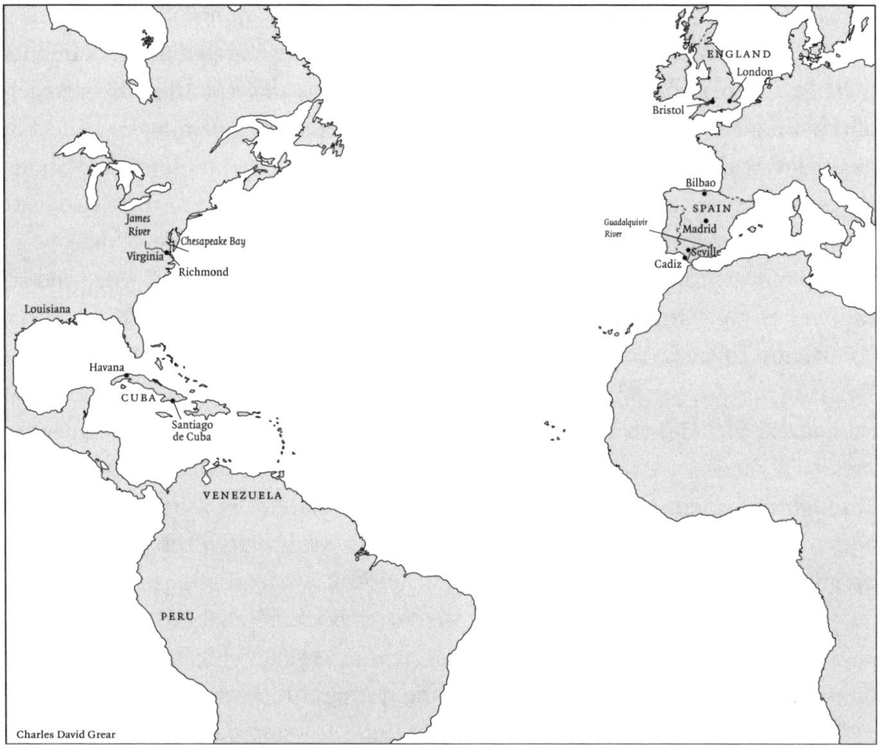

The Atlantic World. Map by Charles David Grear

belongs to the genus *Nicotiana*. Karl von Linné established this genus in 1753, naming it for the sixteenth-century French ambassador to Portugal, Jean Nicot, who was responsible for introducing tobacco to Catherine de Médici in 1560. Domesticated tobacco belongs to two separate species, however. One is *Nicotiana rustica*, which scholars think originated in South America and arrived in the eastern part of North America around AD 160, while the other is the more commonly known *Nicotiana tabacum*, believed to be the tobacco Columbus encountered in the Caribbean.[4]

Tobacco would influence Spain and its colonies—particularly Cuba, along with the broader Atlantic world—in profound ways. This study utilizes what David Armitage describes as a cis-Atlantic approach to examine tobacco and reveal how the plant's use and cultivation affected four areas of life in the Cuba and the wider Atlantic world—culture, society, governmental control, and economics.[5] This work's geographic focus begins in Cuba, where Europeans first noticed the plant, and then moves outward into the Atlantic. Yet, this is not an imperial study of tobacco. In true Atlantic world fashion, forces outside of Havana and Madrid also helped to shape the plant's history. Tobacco's impact on the wider Atlantic world developed because of issues such as the transfer of agricultural knowledge, climate and other natural forces, and disruptions to transatlantic shipping caused by European wars. The story of tobacco stretches far beyond just Cuba, and must include Virginia and Spain, among other locations. This cis-Atlantic approach allows for a more thorough and revisionist analysis of the role tobacco played in shaping Cuba, the larger Spanish empire, and the Atlantic world itself. In short, this study concerns a transatlantic actor, tobacco, within a wider Atlantic context through an emphasis on the location best associated with the plant, Cuba.

Tobacco and tobacco cultivation became linked with the island as Cuba and its capital city grew synonymous with the example of a good smoke. Yet, tobacco's connection to Cuba goes further than just name recognition. In 1775, August Ludwig von Schlözer advocated a multidisciplinary approach to tobacco research that would include, as described by Johannes Wilbert, "religious, therapeutic, medicinal, sociological, economic, commercial, and financial points of view."[6] Through an examination of the interconnected relationships between society, agricultural production, and state control, this work places tobacco within a broader context of Cuba's economic and political development as a colony and an Atlantic actor. A micro- and macro-level study of colonial life on the island emerges, allowing scholars to examine the many ways in which tobacco shaped colonial Cuba, Spain, and the Atlantic world during this time.

Tobacco's impact was not limited to just Cuba or Spain, however, and instead stretched far beyond. While a study of tobacco could address how the plant shaped the world at large, that is beyond the scope of this particular work. The ways in which tobacco affected other parts of the world in terms of culture, society, governmental control, and economics will be addressed briefly in order to illustrate tobacco's role in world history. Tobacco's influence similarly did not end with the colonial period. Instead, it continues to the present day. This work includes examples of tobacco's enduring impact into the twentieth and twenty-first centuries to hint at tobacco's ongoing importance in Cuba, Spain, and the Atlantic world.

If early European explorers and monarchs easily overlooked tobacco's potential, what led to its eventual popularity and spread throughout the world? Chapter 1, "*The consumption . . . is excessive*," explores this question and evaluates the cultural impact of Cuba's most famous weed. The Old World's reception of the plant, especially in terms of how Spain and England regarded tobacco, is of particular interest. These two nations sought to exploit tobacco cultivation soon after its European discovery, with tobacco's popularity driving colonization in certain areas. Yet, many rejected the plant. One of its most famous early opponents was England's King James I (reigned 1603–1625), and tobacco's success as a cash crop for the Virginia colony posed an interesting dilemma for the monarch. As empires expanded and tobacco revenues grew, the plant slowly won over many detractors, including James I.

A form of primitive brand association developed as people began to use tobacco. Tobacco originating from certain locations, most notably Cuba, was highly regarded for its specific taste and qualities. When given the option of choosing tobacco from a particular place, early users stated a preference for the tobacco Spain could provide. Other areas responded to this penchant for Cuban tobacco and adapted their own cultivation and curing methods for tobacco in the hopes of emulating Cuba's product. This work compares practices in Cuba to those in other areas such as Virginia in Chapter 2, "*Learnt by observation and experience*," detailing how agricultural knowledge was transferred. Spain also realized the value of this knowledge as it sought to extend successful tobacco cultivation throughout the empire. The knowledge held by Cuban tobacco farmers was prized and used to teach growers in other locations. This chapter expands the body of work on tobacco cultivation, examining its stages of production from preparing the land for growing to the final stages involving harvesting and curing. Threats to tobacco farmers' livelihoods—from pests, diseases, natural disasters, and other forces—also are examined.

The cultivation and curing of tobacco was not a simple task. "The tobacco farmer, guided simply by the results of years of experience, passed down from his ancestors, knows without being able to explain it, the means to increase or decrease the strength or smoothness of tobacco," a nineteenth-century observer noted.[7] The processes used in Cuba are compared to those used by their Virginia competitors, and establishes the skills needed for successful cultivation, illustrating how knowledge passed north and south, crossing political and economic boundaries.

Who were these people who possessed such valuable agricultural knowledge? Chapter 3, "*One of the good growers*," establishes the background of Cuba's tobacco farmers, also known as *vegueros*, from 1763 to 1817. What characteristics did these individuals share, or were they a diverse snapshot of colonial Cuban society? What types of labor assisted in the production of tobacco? Answers to these questions challenge the prevailing image of the *veguero* established in the historiography of Cuba. Best exemplified in the classic 1940 treatise *Contrapunteo cubano del tabaco y azúcar*, published in English seven years later as *Cuban Counterpoint*, Cuban anthropologist Fernando Ortiz poetically described the island's colonial tobacco farmers, characterizing them as poor, white, Canary Islanders. Subsequent scholars in Cuba and elsewhere continued this description, with many minimizing the role of people of color, slave and free.[8] One scholar even went so far as to argue, "The skill, care and experience necessary for tobacco growing did not allow slave labor."[9]

Yet, research conducted in Spanish and Cuban archives indicate that this is an overly simplistic view of those who grew and sponsored the production of tobacco. Every year in tobacco-growing areas, the *capitán de partido* (district captain or constable) or other another local official collected information and passed it on to monopoly representatives in Havana. These documents, known as *tazmias*, contain valuable information about tobacco production figures as well as the individual growers.[10] Modern scholars can reconstruct a vital aspect of colonial Cuba's tobacco economy through identifying and describing those who were involved in its cultivation. This type of analysis reveals a far more nuanced view of those involved in Cuba's tobacco industry than the one currently presented in the historiography.

Tobacco farmers were not a monolithic group, as period *tazmias* indicate, and therefore Spain's tobacco monopoly led to economic, political, and social effects across many levels of colonial Cuban society. The diversity of the tobacco producers and sponsors meant that the monopoly affected a much larger population than it would have if the *vegueros* had been a monolithic group, which the literature traditionally suggests. Chapter 4, "*A matter of chief importance*," moves tobacco into the international spotlight by detailing how Spain instituted a royal monopoly to capture tobacco's profits for the state. Cuba's 1717 monopoly proved to be the testing ground for tobacco monopolies throughout the rest of the Spanish Empire, a pattern repeated in other Bourbon reforms. Economically, tobacco brought in a sizable profit to Spain, financing numerous activities in the metropole and colonies. This chapter also documents the ways in which the *estanco* (monopoly) controlled colonial Cuba's economy via the tobacco industry. Some of the methods utilized included determining if new *vegas* (tobacco fields) could be established, as well as setting appropriate prices for quality-grade tobacco. Not surprisingly, prices varied throughout the island. Chapter 4 scrutinizes the Cuban tobacco monopoly to determine if it was a success, both in terms of meeting its stated goals, and as a profit-making endeavor.

The impact of this revenue-seeking enterprise went far beyond just economics, however. Cuba's Real Factoría de Tabacos (Royal Tobacco Factory) required a staggering number of bureaucrats throughout Cuba as well as Spain. Chapter 5, "*Apply all of your attention to the laws*," addresses how the tobacco monopoly transformed the political, and ultimately social, structure of colonial Cuba. Tobacco led to the creation of a distinct Cuban society deeply rooted in the new Spanish bureaucratic structures of the monopoly. From the highest-ranking officials in Spain and Cuba to the lowly mule pack driver in Havana, these monopoly workers shaped the island's society. Tobacco was among Cuba's primary crops, yet Spaniards sent directly from the Iberian Peninsula administered the island's monopoly. The increasing administrative presence of Spaniards in Cuba, for example, further cemented the island's tie to the Spanish Crown. The increasing presence of Spaniards in important, and even ordinary, administrative positions within the island of Cuba occurred at a pivotal moment in colonial Latin American history. At a time when many of the mother country's colonies were in full rebellion or newly independent, this was not the case with Cuba. Cuba remained Spain's Ever-

Faithful Isle, both figuratively and literally, during the years when revolutionary chaos surrounded the island. Scholars have addressed the importance of an increasing peninsular population in Cuba during this period, but have not focused on the significance of tobacco bureaucrats as part of this process.[11]

The cultivation and production of tobacco and Spain's control of this valuable commodity had far-reaching labor and social implications in the late eighteenth and early nineteenth centuries. In Cuba's rural hinterland, those who labored in the island's many tobacco fields or sponsored its production reflected the diversity of Cuba's population, as noted earlier. The *vegueros* included, but were not limited to, the slave growing tobacco on a *conuco* (a small plot) and possibly earning enough money to purchase his or her freedom, the Canary Islander, the local priest or retired member of the military, and the titled Cuban noble diversifying his agricultural investments. Yet, all were expected to work within the constraints of the Spanish-imposed royal monopoly to grow and sell their product, no matter their background or social status. This shared reality further linked these individuals, sometimes in defiance of the Crown and its economic policies, and helped to create a society of mutual interests centered on tobacco cultivation and the monopoly that controlled it. This broad opposition base to a key component of Spanish policy should have engendered a culture of open resistance to the monopoly, yet it did not. Rather, these diverse groups turned to more subtle means of resistance, such as tampering with the tobacco delivered to the monopoly or selling tobacco to unauthorized buyers outside of the monopoly process, as detailed in the next chapter. With these alternative mechanisms in place, opposition could take place while remaining hidden, and possibly avoiding detection by royal authorities.

Chapter 6, "*I work with all my might to eradicate the evil*," details the degree to which contraband proved problematic for the monopoly as well as the Spanish Crown. The contraband trade proved pervasive, with strategies such as hiding tobacco, as well as offering it for sale to unofficial buyers, including privateers and pirates. This illegal activity involved *vegueros* and others in the industry, such as the *arrieros* (mule pack drivers) charged with transporting tobacco from the interior of the island to the monopoly in Havana. Such action undermined not only the official tobacco monopoly, but also Spain's hold on the island's commerce. The specific quantity of tobacco in the contraband trade is difficult to establish; however, period documents hint at its magnitude. The numerous and repeated reports sent from Spain, along with those from the captain general to local officials, testify to the extent of the problem and the Crown's futile efforts to stop the illegal trade.[12] As early as the 1750s, Spanish officials decried the contraband trade in tobacco, ordering farmers to deliver their entire product to the royal monopoly houses and prohibiting the sale of tobacco to individuals. In just one year alone, periodic sweeps caught respectable citizens, commoners, mule pack drivers, and even local political officials in the snare. Contraband tobacco could be hidden anywhere. Monopoly representatives recovered tobacco on roads leading to the capital, inside Havana itself, on farms and sugar estates, in homes and on boats. The drive to engage in this illicit trade was so strong that not even tobacco housed inside the monopoly's own mill was safe.

Colonial officials in Spain and Cuba sought to stem the tide of contraband through reforms, hoping to create tighter control over the tobacco monopoly. This chapter compares Spain's response with methods used by other European nations in similar struggles to prevent tobacco contraband. Anti-contraband efforts included use of *guardacostas* (coast guard cutters), the *resguardo* (tobacco monopoly police), and even incentives for those turning in contrabanders. The *vegueros* continued to resist governmental authority and control despite these measures. Yet, contraband was just one means by which the tobacco farmers attempted to circumvent the monopoly's regulations. Another form of resistance was mishandling the tobacco; for example, packing moist tobacco resulted in spoilage. The island's tobacco growers used both overt and covert means to circumvent the monopoly's regulations, and as such, remained a source of opposition to colonial authority until the monopoly was rescinded in 1817.

Tobacco was not immune to forces outside the colony or metropole, however. Transatlantic shipping, and by default the tobacco trade, suffered because of frequent conflicts among European nations, particularly during the latter part of the eighteenth century and the early years of the nineteenth. The Seven Years' War (1756–1763), and revolutionary France's declaration of war against Britain and Holland in 1793 touched the island's royal monopoly, tobacco growers, and Spain's tobacco trade with Cuba. Chapter 7, "*Shipments of Havana tobacco not arriving due to the present war*," details the effects of these incidents. The Iberian Peninsula's tobacco consumption reached such a high level by the late 1770s and 1780s that Spanish tobacco monopoly officials were willing to turn elsewhere in order to meet local demand. In the absence of the preferred tobacco from Cuba, tobacco monopoly and royal officials, even in Cuba itself, consented to the importation of Virginia tobacco in order to meet the ever-growing need. Virginia tobacco was a frequent replacement—albeit a lesser-quality one, according to Spanish officials—when shipping or supply issues prevented Cuban tobacco from reaching the Iberian Peninsula.

The Epilogue brings the story of tobacco and Cuba into the twentieth and twenty-first centuries, examining how the plant continues to shape the wider Atlantic world in terms of culture, society, governmental control, and economics. These issues do not remain in just the colonial past, but are with us today. Tobacco remains a widely consumed product, and one still solidly linked to Cuba. Despite the US economic sanctions imposed on the island since the 1960s, which have recently prohibited the importation of Cuban tobacco products even from third countries, Cuba remains the gold standard among US tobacco consumers for a quality product.

This study seeks to place tobacco and the tobacco industry in Cuba into the wider Atlantic world of the late eighteenth and early nineteenth centuries. Tobacco consumption altered preexisting social constraints and defined leisure time. Cuba's farmers possessed valuable knowledge that was shared with tobacco growers in new areas of the Spanish Empire, in the hopes of replicating the island's success with the plant. Tobacco brought together heterogeneous groups within Cuba dedicated to the cultivation and production of the plant. This group included all levels of colonial Cuban society, from the enslaved African to the retired military member, the free person of color, and the titled Cuban noble. Therefore, the tobacco monopoly's regulations affected a greater swath of the island's population than if the plant had been grown by just a few select segments of Cuban society. The bureaucratic structure created to enforce Spain's tobacco monopoly on the island led to a society with deep Iberian ties. Spain's tobacco monopoly attempted to control every aspect of tobacco on the island, but through various means, including contraband and other passive forms of resistance, the colonists were able to resist. Tobacco was vital to Spain and its users, leading to a populace that demanded access to the product despite wars or other impediments to its shipping. While other locations grew tobacco and could supply the needs of consumers, Cuban tobacco retained a place of prestige in the minds of its users. True aficionados of Cuban tobacco realized this, and little could rival this affection. This was the quality that inspired English writer Rudyard Kipling to opine in his 1922 poem "The Betrothed," "Light me another Cuba—I hold to my first-sworn vows. If Maggie will have no rival, I'll have no Maggie for Spouse!"[13]

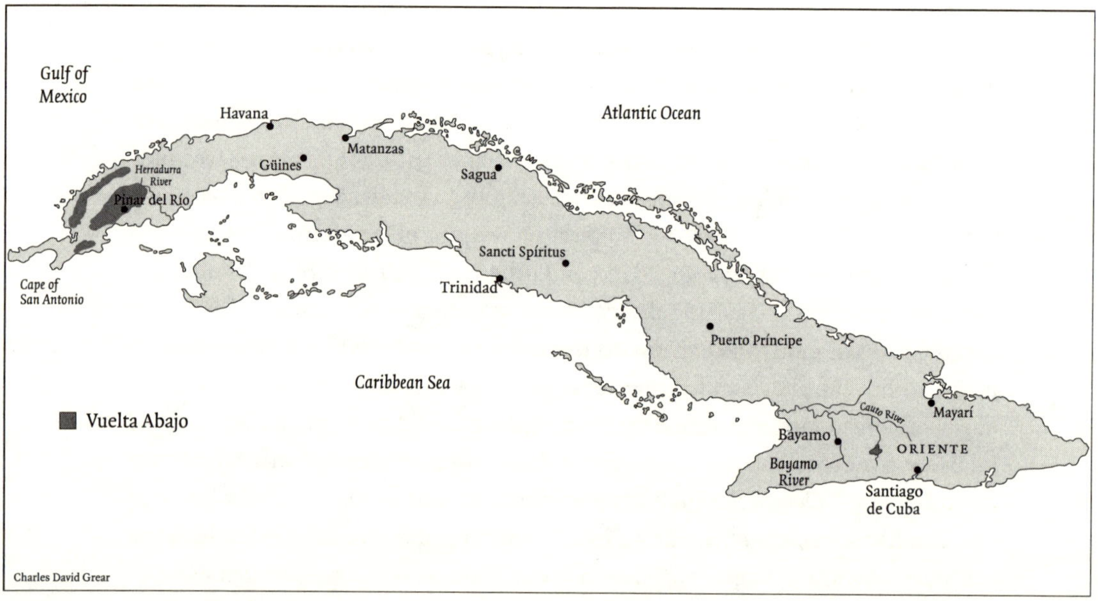

Cuba. Map by Charles David Grear

“The consumption . . . is excessive”

1

SPANISH TOBACCO MONOPOLY OFFICIALS directed Antonio Aguilar de Cela, an inspector at Seville's Royal Tobacco Factory, to head to Havana, Cuba, in early 1778. Aguilar boarded a Spanish frigate, the *Príncipe de Asturias*, in Cádiz. With him were two Spaniards, don Pedro Lopez, aged thirty-four and a resident of Seville, and seventeen-year-old Francisco Miranda from Granada. Aguilar arrived in the island's capital city and began to work at the tobacco factory, the Real Fábrica de Tabacos. The building was located along the bay near the arched *Puerta de la Tierra* (Land Gate), one of the entrances into the walled capital. Contemporary visitors described the tobacco monopoly complex, built by the colonial government during the mid-1770s, as a large quadrangular building with massive, high walls.[1]

One day, as he explored the facility, Aguilar entered an alleyway he believed to be in the heart of the factory. He found himself, to his surprise, inside a hospital where prisoners and other patients lay suffering from a variety of maladies. There he observed tobacco piled “on one side and the other.” Plumbing from the same hospital had “contaminated” the walls of a nearby tobacco monopoly warehouse, causing many *tercios* (bales of tobacco) inside to rot. Even untainted bundles there were unusable, he determined. Aguilar pondered the impact of this discovery, fearing local reaction to “what I have seen with my own eyes.” He soon realized, however, that this “annoyance” would pale next to a possible rejection by consumers elsewhere of Cuban tobacco.

Aguilar wrote in a letter to Cuba's captain general, "What would people in Europe think if they knew about the mixture the tobacco is in that they take with their mouths and their sense of smell?" Aguilar knew the answer and was troubled by it: "The King would absolutely lose this most important branch of the Real Hacienda [Royal Treasury]."[2]

Aguilar was right to be concerned about the economic impact of his unpleasant discovery. Between 1717 and 1817, a state-run monopoly managed tobacco cultivation, purchased the farmers' crop, and controlled the manufacture of its final products. Tobacco was vital to Cuba's economic prosperity, and that of the Spanish Crown. Much of the plant's appeal to tobacco users and producers alike was attributed to what the French describe as *terroir* (characteristics derived from its growing location), a term generally applied to luxury commodities such as wine or champagne. Jean Stubbs argues that the term also applies to tobacco, since it similarly owes its distinct characteristics to the place in which it is cultivated, and to the manufacturing processes it undergoes.[3] If tobacco were to remain a profitable enterprise for Spain, those who ascribed favorable characteristics to tobacco originating in Cuba could never know the conditions in which it was housed on the island. Such revelations could have spelled ruin for the industry, causing untold financial disaster for the island and Spain.

The process that brought tobacco under Spanish control in Cuba had not been an easy one in 1717. The initial reaction was violent. Cuba's tobacco farmers responded in a series of three revolts, including one that saw the captain general flee the island (see Chapter 4). By 1762, Spanish authorities again left Havana by force, this time as the result of a successful British siege during the Seven Years' War (1756–1763). Tobacco remained so important to Cuba and the Spanish Crown that it was a key request in the Spanish articles of capitulation. Spanish officials sought permission to remove tobacco currently in the monopoly's possession, as well as tobacco grown on the island. Although Britain allowed Spanish soldiers to leave the capital city with their "goods, equipages, and money," the successful invaders likewise understood the value of Cuban tobacco, and denied the request.[4] The following year, 1763, Havana returned to Spanish control. While numerous changes in Spanish policy took place because of the war and the Bourbon reforms, the tobacco monopoly remained. Only in 1817, one hundred years after its creation, did Spain allow the free trade of Cuban tobacco.

Much controversy surrounds by whom or when tobacco was introduced to Europe, yet there is little doubt that tobacco originated in the New World. The oldest tobacco seeds of any variety, domesticated or wild, date from 387 to 205 BC and were found near modern-day Tucson, Arizona. Throughout the Americas and the Caribbean, indigenous people knew and used tobacco before the Columbian Encounter, viewing it as a vital element in their social, physical, and spiritual lives. Tobacco was so important to the Navajo people that it figured prominently in their creation beliefs. The Navajo believe that Sky Father and Earth Mother smoked sacred tobacco, and thereby the universe was born. Yet, the Navajo were not the only Amerindian group to hold tobacco in such high regard. In central Mexico, Aztec priests used tobacco in order to communicate with their deities. One particularly Aztec deity, the earth-goddess Cihuacoatl, could even be seen on earth as the tobacco plant.[5]

Tobacco also played an important role in the practical expression of religion and healing. Its predictable and short-lived effects made the plant universally consumed among the New World's Native American populations, which rejected other hallucinogenic plants and herbs whose effects were less reliable. Tobacco's medicinal properties were prized by shamans, who used the plant to diagnose and treat numerous conditions, including asthma, fevers, coughs, rheumatism, headaches, and boils, and to alleviate the pains of childbirth. Five primary methods of tobacco consumption (smoking, chewing, snuffing, drinking, and absorbing via enemas) were utilized in the pre-Columbian Americas. South Americans primarily consumed tobacco in cigar form, while North Americans usually smoked it in a pipe. Although Native Americans employed tobacco in medicine and religion, the plant also served as a leisure product. Spanish chronicler Bernal Díaz, in *The Conquest of New Spain*, noted that Aztec leader Montezuma smoked tobacco following a lavish dinner at his palace.[6] Europeans later used tobacco for similar medicinal and leisure applications, driving its consumption in the Old World and beyond.

Accounts ranging from Columbus's reports to Gonzalo Fernández de Ovideo y Valdés's *La historia general de las Indias* (1535) introduced the plant to Europeans. Sir John Hawkins brought tobacco to England in 1565 after his second voyage to Florida, but the plant's popularity grew slowly and it remained essentially a New World product until the mid-sixteenth century. Nevertheless, tobacco's reach soon extended worldwide, following its introduction to the Philippines (1575), India (around 1595), Japan (circa 1605), Persia (by the early seventeenth century), and West Africa (by 1630).[7]

Tobacco's popularity grew over time in Europe, due in large part to the plant's acceptance by Europeans living in the New World. It was its use as a medicinal herb, however, that was instrumental in its rapid spread. Nicolás Monardes, a Seville physician, was one of the strongest early proponents of tobacco's medical potential. Monardes was the first to make tobacco acceptable not only as a product for European use, but also as a medical resource. His medical history of the Western Hemisphere, published in English as *Joyfull Newes out of the Newfound World* in 1577, detailed how tobacco could provide relief for those suffering from a variety of illnesses and diseases. In his work, Monardes proposed tobacco to treat headaches, help clear the lungs, eliminate worms, and ease toothaches. Translated also into Latin, Italian, French, and Flemish, Monardes's treatise became a staple for European herbalists writing about their practice. As many as sixty separate works from the sixteenth century mention tobacco, with some dedicating a substantial amount of text to the plant. By 1587, an entire book, *De Herba Panacea*, was devoted to the use of tobacco. The impact of these and other printed books contributed to the growing acceptance of tobacco and similar American products.[8]

Yet, not all members of European society found tobacco use acceptable, despite its popularity. The close connection between Native Americans and tobacco resulted in a product considered by some to be uncivilized, and sometimes even demonic. Many in Europe took the attitude that the plant was the "material embodiment of 'bad' Indian culture."[9] Although he touted its medicinal uses, Monardes also suggested that the devil had introduced tobacco to the native peoples of the Americas. Tobacco's acceptance or rejection largely rested on the opinion held by the local ruler. England's Elizabeth I (reigned 1533–1603), an avid smoker herself, approved of tobacco's use in her kingdom, but her successor, King James I, strongly denounced tobacco, writing against it in his *Counterblaste to Tobacco*. In the 1604 work, James I described the practice of smoking tobacco as "a custome lothsome to the eye, hateful to the nose, harmfull to the braine, daungerous to the Lungs, and in the black stinking fume thereof, neerest resembling the horrible Stigian smoke of the pit that is bottomlesse." The following year, he led an even more shocking attack against the habit: sponsoring a public debate on the effects of smoking at Oxford University, he put on display blackened internal organs taken from the bodies of alleged smokers. Other rulers employed more severe measures to persuade tobacco users to abandon the plant, as Jacob Sullum notes. Russia's Czar Michael I (reigned 1613–1645) declared tobacco consumption a crime. By 1643, visitors were reporting the punishments received by tobacco users: whippings, mutilation of their noses or lips, and Siberian exile.[10]

Tobacco faced other powerful opponents throughout the world, including the Catholic Church. Tobacco use, particularly during services, proved to be problematic. In 1624, Pope Urban VIII issued a papal bull threatening excommunication for users of tobacco users, but its use continued, spreading to even the seat of the Catholic Church, St. Peter's in Rome, and forcing Popes Innocent X and Innocent XI to repeat the order. In 1692, five friars paid the ultimate price for their habit. Discovered smoking in the choir during a service, they were sentenced to death.[11] The Catholic Church was not the only religion to disapprove of tobacco use. The Koran did not mention tobacco, which left leading Islamic clerics to disagree about its acceptability. Two Islamic rulers in the Ottoman Empire clearly thought that tobacco use was unacceptable: Sultan Ahmad (reigned 1603–1617) issued bans against tobacco, and Sultan Murad IV (reigned 1623–1640) ordered smokers to face execution. In Iran, Shah Abbas grew particularly frustrated by his soldiers' fondness for tobacco. His punishments of tobacco users included ordering their noses or lips to be cut off, as well as burning some of them at the stake. Royal courtiers who smoked met a nonlethal, yet certainly memorable, fate when he tricked them into smoking ground horse manure. The Puritans in North America also opposed the use of tobacco, noting its addictive and narcotic qualities.[12]

Opposition to tobacco use did not end with the first century after its European discovery. There were members of each successive generation who fought against the ongoing use of tobacco products. By the late nineteenth century, it was no longer doctors' medical treatises or monarchs' pamphlets that decried tobacco. The deaths of prominent tobacco users provided visible examples to those who remained skeptical of the plant's lethal potential. Former US president and Union general Ulysses S. Grant, who increased his habit to twenty-five cigars daily after the 1863 victory at Vicksburg, died from cancer in 1885. His son, General Frederick Dent Grant, also a heavy cigar smoker, died from the same disease in 1916. A longtime pipe smoker, Prussia's Emperor Frederick III succumbed to laryngeal cancer at the age of fifty-six, only ninety-nine days into his reign (1888).[13] However, most doctors, including prominent ones of the day, remained oblivious or ignorant to tobacco's potential dangers. When U. S. Grant developed mouth cancer, his physicians cut his intake of cigars to three per day, but did not eliminate them. Sir William Olser's 1892 treatise, *The Principles and Practices of Medicine*, contained only three sentences about tobacco's effects within its one thousand pages.[14]

Despite all its dire health effects, tobacco would have played but a small role in Cuba and the Atlantic world if it was not popular. Tobacco users began to develop a primitive brand association and loyalty soon after the first tobacco shipment from the New World to the Old in 1598. Even in this early period of tobacco consumption, users indicated a preference for one type of tobacco over another. The English favored tobacco from Spanish America, consuming an estimated 130,000 pounds of Spanish tobacco worth two hundred thousand sterling annually as early as 1605. English imperial powers sought to give a privileged status to Virginia tobacco, yet Cuban tobacco remained a favorite. Despite higher prices, "throughout the seventeenth century, Cuban tobacco gave evidence of being far superior to that of Virginia and English smokers were happy to pay the additional price."[15] England nevertheless sought to compete with Spain's product, beginning with the production of tobacco in its own New World colonies.

John Rolfe introduced tobacco to Virginia in 1612, and Virginia's first shipment of the plant entered the English market the following year. Pressure from Virginia interests and favorable taxation practices allowed British colonial tobacco to shift market preferences away from foreign tobacco. Virginia tobacco enjoyed a 600-percent preference in customs duties, and severely challenged Spanish tobacco imports, as reflected in Table 1.1. An upsurge of tobacco coming to Britain from its North American colony contrasted with a drop in Spanish imports to

TABLE 1.1: Tobacco imports to Europe, seventeenth and eighteenth centuries

YEAR	FROM	TO	AMOUNT (IN POUNDS)
1609	—	Seville, Spain	15,328
1613	—	Seville, Spain	404,564
1615	Virginia	England	1,250
1615 and 1616	Spanish America	England	50,000
1618	Bermuda	—	30,000
1618	Virginia	—	24,834
1620	Virginia	—	40,000
1620	Bermuda	England	17,000
1623	Bermuda	—	80,000
1623	Virginia	—	75,864
1630	Virginia	England	370,000
1638	Virginia	—	1,400,000
1638	Barbados and St. Kitts	England	675,000
1639	Maryland	England	100,000
1640	Barbados and St. Kitts	England	200,000
1664	Virginia	England	24,000,000
1790	Virginia	—	72,000,000
1796	Virginia	—	34,000,000

Note: Locations indicated by a dash (—) are unspecified in the sources.

Sources: Axton, *Tobacco and Kentucky*, 25; Wesley Frank Craven, "An Introduction to the History of Bermuda," *William and Mary Quarterly*, Second Series, 17:3 (July 1937), 356; Goodman, *Tobacco in History*, 139, 143; Levi Marrero, *El siglo XVII (II)*, volume 4 of *Cuba: economía y sociedad*, 39-40; Laura Nater, "Colonial Tobacco: Key Commodity of the Spanish Empire, 1500-1800," in *From Silver to Cocaine: Latin American Commodity Chains and the Building of the World Economy, 1500-2000*, eds. Steven Topik, Carlos Marichal, and Zephyr Frank, (Durham, NC and London: Duke University Press, 2006), 95; Ravenholt, "Tobacco's Global Death March," 216; Kulikoff, *Tobacco and Slaves*, 157-158; Michael J. Jarvis, *In the Eye of All Trade: Bermuda, Bermudians, and the Maritime Atlantic World, 1680-1783* (Chapel Hill, NC: University of North Carolina Press), 27.

England, which declined to fifty thousand pounds annually just a few years later. By the 1620s, tobacco imports from Virginia, which averaged sixty-five thousand pounds, surpassed those from Spain and Portugal. Virginia's tobacco exports to England rose to twenty-four million pounds annually by 1664, a figure that only grew as Maryland, Pennsylvania, and the Carolinas initiated their own tobacco production.[16] Approximately fifty years after its first introduction, annual tobacco consumption per capita in England and Wales increased from 0.01 pound in the 1620s to 1.10 pounds in 1672. Per capita tobacco consumption reached 2.30 pounds by the turn of the eighteenth century, a staggering increase of 22,900 percent in less than one hundred years.[17]

Spain lost its foothold on the English market in 1619 as Virginia's tobacco imports to England exceeded those from the peninsula, yet Spain's own citizens eagerly filled the gap. Tobacco imports to Seville exceeded seventy thousand pounds by 1608, a dramatic increase of 2,519 percent over imports just six years earlier.[18] Increasing demand for tobacco products led to the creation of the first tobacco facility in Seville in 1620. The critical moment, however, was in 1684, when a royal order gave the factory official backing as the headquarters of the *renta de tabacos* (tobacco monopoly). The directive also gave the factory the right to produce all tobacco products destined for consumption within Spain. The San Pedro facility expanded in physical size several times over the next half century in response to rising consumption, yet plans to construct a new tobacco factory near the banks of the Guadalquivir River were already underway by 1725. This move could not come soon enough, as Spain's tobacco consumption only continued, rising by a little more than 8 percent in six years from 2,667,000 pounds in 1749 to 2,883,000 pounds in 1755.[19]

Travelers throughout the nineteenth century areas who brought home new and different ways to consume tobacco shaped consumption habits in their home countries. Soldiers returning to Britain from the Napoleonic campaigns in the Iberian Peninsula (1808–1814) introduced cigarettes, while British soldiers home from the Crimean War (1853–1856) fostered a marked increase in cigarette smoking among the British Isles. Tourists headed west across the Atlantic introduced the custom to the United States shortly thereafter, and the practice of smoking cigarettes grew during the US Civil War (1861–1865). Chewing tobacco, however, remained the United States's primary method and form of tobacco consumption throughout the nineteenth century, although cigarette smoking continued to grow in popularity.[20]

As tobacco production and consumption spread, tobacco began to shape early modern society in powerful ways. Production specialization soon after tobacco's introduction to the Old World led to an increasingly selective clientele, one who recognized the differences between tobacco grown in particular locations and under certain conditions. Not unlike the qualities of *terroir* described by vintners and wine connoisseurs, this fostered an early form of brand association. Tobacco growers and their national sponsors promoted the qualities of their tobacco in order to increase sales and protect their individual markets. Cuban tobacco in particular earned a reputation as an excellent product, preferred by many tobacco users. The special qualities present on the island and the characteristics it imparted to the tobacco grown and manufactured there led to a quality product not replicable elsewhere. An unnamed author, described by the *Chicago Daily Tribune* as "well acquainted with the agricultural products of Cuba," wrote in late 1898 that "the climate of Havana offers conditions that enable the crude tobacco product here to be manufactured under the most favorable circumstances. The air here furnishes just the right amount of moisture and the proper amount of heat; in fact, it is said by experts to furnish everything necessary. At other places these conditions may be supplied artificially, but here the leaf slowly and from entirely natural causes reaches the stage where it is to be made into cigars."[21]

Tobacco producers living outside of the island, of course, did not have access to the particular Cuban soil or climatic conditions. However, they too found a way to produce what they argued was comparable tobacco by using Cuban tobacco seeds. Texas growers near Gonzalez and San Marcos experimented with Cuban seeds as early as 1852, in the hope of developing its production there. By 1892, the Dallas County exhibit of its agricultural products at the Texas State Fair included a "statue of King Tobacco" constructed entirely of tobacco grown from Cuban seed.[22] California growers likewise offered the possibility of access to Cuban tobacco cultivated in the United States. A Mr. J. R. Nickerson, originally from Kentucky, claimed at the 1860 Marysville District Fair that California could raise tobacco similar in quality to that elsewhere in the Union.[23] Tobacco users, discerning as they may be, were warned against trusting that all tobacco from the island was the same, however. An 1883 article in the *Columbus (OH) Daily Enquirer-Sun* warned smokers that not every imported Cuban cigar was the prized "Havana." Very little tobacco from the island warranted that title, and what did usually was consumed within Cuba itself, or sent to "satisfy the aesthetic demands of the royalty and aristocracy of Europe."[24]

Ultimately, however, the desire to consume tobacco usually overrode preferences when product from a favored location was unavailable. Tobacco consumption in Spain reached such a high level by the late 1770s and 1780s that Spanish tobacco monopoly officials were willing to turn to an inferior product in order to meet local need, citing the "excessive" consumption on the peninsula. Monopoly and royal officials even in Cuba by the early 1800s had conceded to the importation of Virginia tobacco in order to meet the ever-growing demand on the island and at home in Spain. Spanish consumers favored Cuban tobacco, but when pressed, they were willing to get their tobacco from anywhere. Transatlantic shipping suffered because of frequent conflicts among European nations, chiefly during the latter part of the eighteenth century and the early part of the nineteenth. None other than Spain's minister of finance, Miguel Cayetano Soler, realized the challenge this posed for Spain. "Every day the [tobacco] crop has been decreasing, as well as the quality of the leaf and ground tobacco; Spain sees itself deprived of this aid and in the difficult position of availing itself of the disagreeable Virginia tobacco to support in the method possible its consumption and avoiding the ruin of this precious income," he wrote in 1804. This "disagreeable" Virginia tobacco was a frequent replacement, albeit a lesser quality one, when shipping or supply issues prevented Cuban tobacco from reaching the Iberian Peninsula.[25]

Tobacco revealed other changes in early modern society. All sectors of society consumed tobacco, uniting rich and poor in a pattern of collective addiction as it shaped early modern daily life. Tobacco reached across socioeconomic levels at about the same time, unlike sugar, chocolate, or tea, which were introduced to the elites and eventually made their way down to the masses. In his seminal study, *Sweetness and Power*, anthropologist Sidney W. Mintz notes that sugar remained essentially an elite consumer product until the eighteenth century. As with other spices, sugar was a luxury reserved for those with power and economic means. Tobacco's popularity, in contrast, resulted in expanded production, quickly leading to a drop in prices, making the new commodity available to nearly all segments of society as early as the 1620s and 1630s. As tobacco use among Europeans became more widespread and accepted, it assumed a central place within Atlantic society and culture. Tobacco became such an ingrained part of the local culture and society for Spain and Cuba that it easily crossed all social and class boundaries.[26] A resident of Madrid in 1634 revealed how pervasive tobacco consumption was in the Spanish city. Tobacco was used by "the soldier to the student, from the monk to the priest, from the city-dweller to the rustic, from the plebian to the noble, from the youth to the old man, there is hardly anyone who had not at least tried it, and even more use it habitually," the writer noted.[27] By the early 1700s, as a result, Spain's tobacco consumption exceeded 3.5 million pounds per year.[28]

Numerous foreign visitors to Cuba in the mid-nineteenth century noted the centrality of the plant to the island's life and culture. Irish doctor Richard R. Madden commented on tobacco's ability to unite all ages, social classes, and racial classes in his 1849 account: "The priest, the lawyer, the doctor, the businessman, the landowner and his juniors, the women (with the exception of the ladies of the highest rank), children from age seven or ten years of age, city slaves, and the free people of color . . . smoke and make smoking a serious business with between ten and up to twelve cigars [*tabacos*] a day." Just ten years later, little had changed when Joseph J. Dimock visited the island. "It *seems* as if every man, woman, and child smoked. . . . The men of all degree smoke, and smoke everywhere: in the streets, in the houses, in cafes, theatres, counting rooms and while eating and drinking. The shoemaker over his lapstone and the tailor over his goose, all smoke, smoke and the first courtesy offered you by a Cuban is the everlasting cigar," Dimock recounted.[29]

What makes this remarkable is not that so many people in Cuba of all different occupations participated in tobacco consumption but that through doing so they frequently crossed otherwise taboo social boundaries of gender, race, and class. The necessary activities involved in tobacco consumption allowed members of disparate groups to interact with one another in ways that were socially exceptional. American lawyer Richard Henry Dana Jr. traveled to Cuba just before the US Civil War; he commented, "The cigar is a great leveler. Any man may stop another for a light. I have seen the poor porters, on the wharf, bow to gentlemen, strangers to them, and hold out a cigar, and the gentlemen stop, give a light, and go on—all as a matter of course." Yet, tobacco was more than just a communally addictive pastime in the island's otherwise strict social setting. Colonial Cuban society operated under very rigid social codes, ones that isolated even further foreigners and women. Luis Martínez-Fernández notes this social stratification, arguing, "Clearly established social rules designed both to 'protect' and subdue women contributed to keeping white Habaneras under seclusion." Even stronger than the social boundaries associated with gender were those linked to race. As Martínez-Fernández further explains, "A slave-based society, Cuba's complex social structure rested, above all else, on status: people were either free or slave. There was, to be sure, a close correlation between race and status."[30] These types of interactions between such dissimilar levels of colonial Cuban society would rarely have occurred, particularly at such an intimate level. Yet, over tobacco, Cubans temporarily were willing to put aside such social constraints.

This accessibility of tobacco across socioeconomic borders distinguished it from its use by some pre-Columbian Native American groups. Those like the Aztec and others considered it an exclusive product for those deserving of the herb. For the Aztec, tobacco played an important role in special ceremonies that celebrated their social hierarchy system, which was based on "bloodlines and battlefield virtuosity." Aztec warriors received tobacco pipes and chocolate beverages as rewards for their actions.[31] The very visible presence of tobacco in ceremonies centered on key elements of Aztec society, warfare, and social stratification demonstrate the importance of tobacco as an elite product to this Native American group, unlike its use by Europeans, who viewed the plant as a more egalitarian vice.

While tobacco consumption itself crossed social and economic boundaries, the early European methods of consumption frequently revealed social status. Pipe smoking during the 1700s, for example, remained a vice of the lower classes, while those of higher means turned to snuffing, a practice developed in France. Even into the nineteenth century, pipe-smoking patterns reflected social standing. Those of higher social standing preferred "the lighter and more finely cut Virginian flake tobaccos, which were ready to smoke," while lowly Welsh miners chose "strong shag tobaccos" of coarse-cut leaf and rolls. Dockworkers favored thick twists of tobacco, while cabmen chose "Irish roll."[32] These social distinctions in tobacco use were not limited to Europe, however. In the early colonial era, tobacco use among the British North American colonists also varied by location and occupation. Urban areas tended to favor snuff, while chewing tobacco was seen as a habit reserved for sailors and those who worked in certain physical labors.[33] Over time, tobacco consumption methods centered around smoking in all forms and across all social classes.

Tobacco consumption also transformed modern society by altering contemporary views of leisure. In his study, Peter Borsay defines leisure as existing outside of the necessary activities of daily life such as eating and sleeping. Perhaps, equally important, Borsay argues, is defining leisure in contrast to what it is not—work.[34] Tobacco consumption moved leisure from an activity enjoyed exclusively by the elite to one more collectively practiced, instead becoming a pursuit experienced by the rich and poor, as well as men and women. With the advent of the modern age, leisure now entered more social levels. Those of economic means had always enjoyed free time for hobbies and other pursuits, but the working class had not possessed the same advantage. Work and the business of living had consumed most, if not all, of their time. The Industrial Revolution changed that. Mechanization brought at least brief periods of rest for the worker and tobacco consumption became a more universally enjoyed activity. Yet, as Borsay notes, leisure did not need to be an undivided activity. Periods of undivided rest were difficult to find for those in the middle and working classes. These groups in response increasingly utilized periods of semi-relaxation.[35] Tobacco use was a vital component in this type of semi-leisure since it could be enjoyed while doing other activities. Such temporary and shared periods of rest nevertheless helped to alleviate the stresses and pressures of peasant and middle-class life, all while one continued to go about other necessary activities.

One manner in which the working class and poor used tobacco was as a source of solace and comfort. David T. Courtwright argues that the initial success of "psychoactive substances" in early modern Europe was their ability to help "peasants and workers cope with lives lived on the verge of the unlivable."[36] This is one area in which tobacco's hunger-suppressing qualities may have made the habit additionally appealing. An Elizabethan-era doctor noted this particular quality as early as 1602 in a treatise against tobacco consumption.[37] Even as late as 1823, revenue commissioners in Ireland noted that lower-class men and women alike were the primary users of tobacco, concluding that "as a question of feeling it would be a good thing to give it them cheap, it is such a source of comfort to them."[38]

Yet, as Peter Burke argues in his 1995 article on leisure in early modern Europe, leisure existed primarily as a male right and experience. Tobacco consumption remained socially acceptable in many circles only for men for many years. Women, at least those of respectable social standing, did not partake in the habit. Victorian-era associations between tobacco consumption and women deemed of questionable character, such as actresses, prostitutes, and gypsies, were common. The connection between Seville's famed tobacco factory and women of low repute extended even to those producing tobacco. Local Spanish lore has it that statue of Fame above the main doorway of Seville's factory will not blow her trumpet until a virgin enters the building. Similarly, nineteenth-century writer Prosper Mérimée and composer Georges Bizet both depicted the character "Carmen" in a sexualized manner as a shameless Seville factory worker, rolling cigars on her bare thighs.[39] Descriptions from contemporary visitors to the Seville factory in the 1860s only perpetuated this image. One observer reported that each woman was "dressed only in a chemise and a skirt. . . . Many of them ignore the use of stockings." While most of the women wore "crinolines, or cages" in the street, "they took them off and hung them up around the pillars of the rooms, with their shawls, their *mantillas de tira*, and the baskets which contained their repasts" when in the factory.[40] In *Vistas de España*, Elizabeth Boone points out that "in [illustrator Gustave] Doré's drawing, the *cigarrera* at left reveals a tempting bit of neck and shoulder; her unshod foot, bare heel and ankle, and a portion of her calf peek out from the hand-patched skirt of her loose-fitting dress."[41]

Whether it was revered or reviled, Cuban tobacco helped to shape consumption habits throughout the world. A Cuban tobacco product became the gold standard for growers worldwide, and they attempted to emulate the plant's success, even importing Cuban tobacco seed. Like sugar, coffee, tea, and other products, tobacco influenced world health and workers' labor patterns, as well as transatlantic trade and colonial development. The process necessary to transform the minute seeds to ground, snuff, cigarette, or cigar products was a complex procedure. The growing, harvesting, and curing of tobacco required a specific skill set and knowledge that was shared north and south as farmers from Cuba to Virginia to Peru sought to learn from one another.

2 "Learnt by observation and experience"

THE CLOSE CONNECTION BETWEEN CUBA and tobacco in product recognition and preference influenced the already booming tobacco consuming habits of the Atlantic world. Tobacco cultivation in particular helped to shape the development of colonial Cuban agricultural labor practices. Farmers needed specialized knowledge to yield a quality tobacco crop, adapting to local and climatic conditions. Likewise, those who processed tobacco into the final consumer product also possessed particular skills. Two disparate locations, Cuba and Virginia, each known for its cultivation of tobacco during the colonial period, utilized many of these same techniques and practices. Faced with a lowered supply of leaf, Spanish colonial tobacco officials even turned to Virginia in order to supplement their tobacco needs. Between 1779 and 1783, in 1799, throughout the first two decades of the nineteenth century, and even into the 1860s, this practice of purchasing Virginia tobacco continued. Spanish officials found the replacement an unequal substitute, but noted that the need for tobacco left them little choice.[1]

Tobacco cultivation, harvesting, and curing involved many different procedures and steps, but farmers in Cuba and Virginia during the eighteenth and nineteenth centuries often followed similar practices. Farmers in Cuba and their counterparts in Virginia generally agreed on what constituted the proper way to grow and harvest tobacco. From the land best suited to the cultivation of tobacco to the preparation of the seedbeds, and the method of harvesting the tobacco leaves, contemporary instructions for tobacco farmers in Cuba and Virginia tobacco farmers were remarkably similar. Even before a colonial government could regulate tobacco sales, or the plant enjoyed in its various forms, it had to be planted, tended, cut, cured, and processed. Many stages from field to final product separated tobacco growers and users, yet the two shared a deep-seated kinship; according to scholar Jordan Goodman, this prompted tobacco farmers to use language similar to that employed by the plant's consumers.[2]

Tobacco cultivation was a particularly difficult process in which climate, soil, seed, and the method of culture all combined to determine the crop's growth and ultimate quality. An experienced grower knew when to transplant the young tobacco plants, or when to cut it, or how long to leave tobacco to cure, experience that only served to improve his chances of success. The amount of nicotine, essential oils, and other elements found in the plant, for example, varied depending on when it was cut. Farmers also had to take into consideration numerous other factors such as final color, flavor, and aroma in order to ensure a quality product. Together, all these qualities contributed to the particular *terroir* enjoyed by tobacco.[3] Yet, the knowledge held by farmers was a key component in bringing out a fine tobacco's true potential, prompting a nineteenth-century Cuban *veguero* to remark, "The planter who babies his tobacco most is the one who gets the best crop."[4]

The numerous steps of bringing the product to market set the pace of life in colonial Cuba and Virginia. Tobacco cultivation required little in the way of physical capital, but depended heavily on hand work. Unlike wheat or sugar, there were few slack periods for the tobacco farmer. The farming followed a strict schedule, dictated by the crop itself, and one in which every activity was equally important. Yet, not all the nineteenth-century visitors to Cuba appreciated the work that went into bringing tobacco to the market. American William Morton wrote in his diary, "Their tobacco and corn is planted very thick and gets very little work. The people are generally lazy and do as little as possible."[5]

Skilled Cuban *vegueros* possessed knowledge about how to produce a successful tobacco crop. When Spanish officials wanted to improve tobacco cultivation in Peru, they turned to these same farmers. On 28 June 1811, a royal order was issued to the head of Cuba's tobacco monopoly, requesting that four tobacco farmers be sent to Peru with four *arrobas* (one hundred pounds) of tobacco seed. A public announcement distributed throughout the island that fall sought volunteers. Once in Peru, *vegueros* were informed that they were to "establish the cultivation in that kingdom, and teach to those natives the method of this cultivation: the subjects for consideration should have the aptitude necessary for the object indicated, and will present themselves in the Secretary of the Superintendent's office to be accredited, and instructed of the conditions of verification for the job."[6] By mid-century, Cuba remained the site for sought-after training in tobacco cultivation and processing. An 1859 request to the Spanish minister of war and overseas affairs demonstrates this vividly. It was suggested that ten or twelve soldiers from the Canary Islands' provincial corps be sent to Cuba for a year in order to "study the cultivation and preparation of tobacco," with the government paying for their trip and maintenance while on the Caribbean island. Those familiar with Cuba's history and the mythology surrounding tobacco may have found the 1859 request ironic, as Canary Islanders in Cuba had long been viewed as expert tobacco growers. This would not be the last time that Cuban *vegueros* and the methods they used to cultivate tobacco served as examples for others in the Spanish Empire. A manual for Venezuelan tobacco farmers published in 1883, for example, stated that its methods were based on those implemented in Cuba, and adapted for Venezuela.[7]

The ever-growing market for tobacco meant that growers throughout the world were in direct competition with each other. Spanish authorities made a point of staying informed about their competitors' product, but they ultimately wanted to produce the best goods possible and were willing to learn from others in order to do so. This included admitting when tobacco growers in other locations created a product that they could emulate. Spain's minister of finance, Miguel de Múzquiz, in the mid-1770s alerted the Marqués de la Torre (Felipe de Fondesviela), Cuba's captain general, of a special package that would arrive soon. Spain was sending two small bundles of Brazilian tobacco with the mail ship, and asked that Cuba imitate the "shape, thickness, [and] turned point" of the provided samples.[8] Yet, officials and tobacco users knew that not all tobacco was equal. Tobacco purchasers favored an exceptional product, making quality of the utmost importance. Each step in bringing the final product to the consumer, from seedbed to factory floor, was critical in developing a *terroir* favored by customers. Possible ruin awaited in various forms, from pests to unexpected storms, or even transportation difficulties and shipping delays caused by European wars. Taking all these factors into consideration, if a quality tobacco product reached the hands of customers, it was a matter of luck, rather than certainty. As scholar Barbara Hahn effectively argues, much of the distinctiveness of Cuban or Virginia tobacco lay not in the particular variety or seeds, but rather in the "impact of cultivation methods on plant processes."[9]

One of the most vital factors affecting the quality of tobacco was the choice of land and its preparation for cultivation. Eighteenth- and nineteenth-century tobacco farmers in Cuba, along with those in Virginia, agreed that a slightly sandy soil was best for tobacco cultivation. Land along rivers or a stream generally was deemed ideal for Cuban and Virginian farmers alike. One contemporary

Cuban writer, Martín Fernández Miñano, disagreed. He recommended hill plots, noting that higher ground was less prone to flooding. This would prove to be good advice, as the historical record occasionally indicates damage to tobacco fields because of flooding. Most farmers agreed about the type of soil necessary to produce a quality product, however. The correct way to prepare property for cultivation varied greatly according to the Cuban and Virginian tobacco manuals. De la Sagra's 1839 Cuban manual and Peter Minor's 1822 instructions for Virginia farmers both advocated burning any vegetation standing on the parcel in preparation for planting. By 1874, however, J. F. Nobiatur advised Cuban *vegueros* not to burn the land before planting.[10] This changing view of the effectiveness of fire in land clearing may have simply been a result of a reevaluation of its effectiveness by this later period. A practice utilized by North American indigenous populations prior to and after the Columbian encounter, the burning of dead wood and live vegetation grew out of favor over time. Seen as "wasteful and destructive" by the early 1900s, this practice added nitrogen-rich ash to the soil, which increased fertility in the coastal plains, although it contributed to damaged humus and eventually led to soil erosion in rocky, rolling areas.[11]

Over time, finding new tracts for tobacco cultivation became an issue for some planters, particularly those from Virginia. They argued that tobacco exhausted the soil, requiring them to rotate crops every couple of years and to leave the land fallow. In some instances, land planted in tobacco for three years needed twenty years to recover. To continue quality production of tobacco, farmers who did not already own a sizable amount of land were forced to move into new areas for cultivation. This meant that new land continually was sought for Virginia's tobacco fields. Settlement and tobacco cultivation began in the Tidewater area near the Chesapeake Bay and spread westward. By 1785, settlers had pushed into what would later become the state of Kentucky and began to grow tobacco there. Those remaining in Virginia and Maryland, the traditional eastern areas of tobacco cultivation, faced another hurdle—expanding population and limited land.[12] One scholar notes that by 1800, the growing population of Virginia and Maryland meant that enough choice land was no longer available. With a potential labor force of 383,333, and 6,820,200 acres of prime land, each worker had less than eighteen acres—"less than half the fifty acres considered necessary."[13]

These concerns about soil exhaustion appear to be limited to farmers in the United States, however, as Cuban farmers did not seem to pay much attention to such matters. Instruction manuals and period correspondence from Cuban tobacco monopoly officials indicate that late eighteenth-century and early nineteenth-century farmers, at best, showed limited concern with tobacco's impact on soil fertility and little interest in crop rotation. One lone example has been found to date. In April 1766, Trinidad's *cabildo* (town council) authorized local *vegueros* to keep the lowest class of tobacco in order to benefit the soil. Cubans additionally may have shared the views of a Virginia planter who argued in 1822 that tobacco did not exhaust the soil:

> Tobacco has been reproached as the canse [*sic*] of the general exhausted condition of our lands, of the slow paced improvement in the Virginia system of agriculture; in short as the bane of all good husbandry. This stigma is, I am persuaded, in a great measure unmerited. It is true, that like Indian corn, from the frequent and high degree of tillage it requires throughout the summer, it exposes the ground to be washed by hard rains, and evaporated by the hot sun; but the plant in itself is less an exhauster than corn or wheat. A proof of this is to be found in the superior growth and perfection to which any crop will arrive when grown after tobacco, than after anything else, not excepting clover that has been ploughed in.[14]

Perhaps this attitude was due to the limited amount of land available to Cuba's farmers. As an island, there was only so much land available, and over time, demands from other agricultural pursuits meant that *vegueros* were lucky to have access to the land they did. To consider moving farther and farther inland to find virgin land was a moot point. This was not the case in Mexico, however. Tobacco monopoly bureaucrats there criticized what they viewed as the apparent ignorance of the planters. There, not only did Mexican officials reproach the farmers for failing to improve worn-out soil, but the officials also claimed that the tobacco planters insisted on producing "tobacco on any land whether good or bad."[15]

Cuban and Virginian instructions to tobacco farmers agreed that slightly sandy soil was best suited for tobacco fields. The proper location for seedbeds or when to sow them was a matter of dispute. Peter Minor recommended that Virginia tobacco farmers choose land at the foot of a hill, preferably located near water or a stream. Several nineteenth-century Cuban instructions advised the use of virgin land for tobacco seedbeds, a practice conducted since "time immortal."[16] Similarly, the best time to sow tobacco seeds in the seedbed was a great source of disagreement among Virginian and Cuban *vegueros*, mostly because of the local climate. While manuals advised Cuban *vegueros* to sow their seeds periodically beginning in October and ending in December, the weather in Virginia was far too harsh to start the tobacco seeds at this time. Minor recommended instead that the farmers begin sowing in early February. Favorable weather conditions in Cuba, however, allowed farmers to plant a second seedbed in April.[17]

Tobacco seeds were extremely diminutive in size, with one ounce containing approximately three hundred thousand seeds. A single tobacco plant left to flower and produce seed was capable of supplying half an ounce of seeds, sufficient to plant one hundred square yards of seedbed. Ideally, seeds from one plant could yield enough seedlings to plant between two and five acres of tobacco.[18] Put in other terms, one tablespoon of seeds could plant six and one half acres of land in tobacco. Farmers mixed tobacco seeds with fine sand to facilitate distribution in a manner similar to that of sowing wheat, and scattered them over the seedbed, a process noted in a 1788 Cuban manual.[19] In Cuba, a palm broom was used to cover the seeds with a light layer of dirt. Cuban farmers protected the seedbeds with green plantain leaves to prevent damage by rain or sun. At night, however, farmers removed the covering to allow the young plants to get moisture from dew. Extremely cold weather did not pose a problem in Cuba. As a result, farmers did not need to remove the brush covering the seedbed in March before the plants came up. Nor was it necessary for them to re-trample the ground, add a dressing of sifted fowl manure, and re-cover the seedbed, as farmers in Virginia were advised to do in case of severe frost.[20]

Growing tobacco was a multistep process. Land was prepared for a seedbed, the tiny seeds were sown, and the young tobacco plants later transplanted to their final place in the field. This period of several months was a critical time for the tobacco grower. The farmer routinely grew many more seedlings than he needed for transplanting. A 1788 Cuban document, for example, recommended a ratio of three seedlings to one final plant. If any young plants were damaged before or after transplanting, additional seedlings were available. The hope was that in this way the farmer could offset any possible small-scale loss of his crop.[21] Yet, any injury to the seedbed meant a potential loss of the entire year's crop, even before it reached the field. Tobacco was especially susceptible to damage from flooding. Standing in water for more than a day caused the plant's roots to die, a condition only hastened in warm soil like that found in Cuba. Damage to the entire root system signified permanent wilting and a ruined plant, while the destruction of only some of the deeper roots resulted in a temporary wilting of the leaves.[22]

The tobacco-growing area of Güines, Cuba, was particularly hard hit in the fall of 1768 by a hurricane in mid-October, followed by severe flooding in early November. Captain Francisco Gama noted that the river "rose like never we have experienced," sending water into "many seedbeds" and destroying the young plants there.[23] A closer examination of Captain Gama's report suggests that water damaged the entire root system by not receding quickly enough to allow the soil to dry in order to save the plant, resulting in a total loss of the crop.

Tobacco plants were ready for transplanting to their final place in the field when they had four true leaves, which occurred anywhere from two to three months after the seeds were first sown. Cuban tobacco farmers were advised to remove the first crop of young tobacco plants from the seedbeds between December and March, preferably from three in the afternoon onward. For Cuban *vegueros*, the ideal was to place the rows approximately one *vara* (thirty-three inches) apart, with the distance between the plants themselves to be *media vara* (approximately sixteen to seventeen inches). One *caballería* (approximately thirty-three acres) contained an average of three hundred thousand plants once they were established in the field. Tobacco plantation owners around the Chesapeake advocated a slightly wider distance between plants of about three feet, resulting in an acre of tobacco containing only several thousand plants.[24] Thus, a Cuban planter, with his closer spacing of tobacco, was able to harvest a larger crop on the same amount of acreage as compared to a planter in Virginia.

While the tobacco plants grew in the field, the farmer's work continued unabated. Dealing with pests, addressing tobacco diseases, controlling weeds by regular hoeing (with an iron hoe that weighed between eight and ten pounds), removing dead plants and replacing them with new ones, turning leaves to face upward after a period of high winds, and keeping the tobacco plants watered were all tasks that occupied his time. Three other important operations—priming, topping, and suckering—aided in the development of a strong, productive tobacco plant. Farmers removed several leaves at the bottom of the plant as part of the priming process. These leaves, in contact with the ground, were often damaged by rains or hoeing and, as a result, were inferior in grade to those higher on the plant. Removing these leaves allowed the plant to use its energy to produce higher-quality leaves, in addition hopefully to slowing the access of pests and diseases to the plant. Approximately two months after the tobacco was transplanted, between February and May for Cuba's first tobacco crop, or from August to September for the second crop, the grower engaged in a practice known as "topping" or "debuttoning." Here he made his way through the fields early in the morning, pinching off with his thumbnail the top of the plant, including the newly formed cluster of pink flowers and several leaves. This allowed the plant to concentrate its energy on the leaf development rather than seed production. It was during this particular operation that the expertise of the Cuban *veguero* became evident. The use of unskilled labor in this vital operation could be disastrous, as it was essential that the process maintain the appropriate number of leaves on each plant. Topping the tobacco prompted the growth of offshoots at the base of the plant. Called "suckers" in Virginia and "hijos" (children) in Cuba, they too had to be removed lest the plant's overall leaf quality suffer.[25]

A *veguero*'s knowledge certainly helped to ensure his success in bringing the crop to market, yet even the most seasoned farmer could do little when confronted by an uncooperative Mother Nature. The quality of the soil or local climate conditions sometimes meant that even the most experienced planter—or a future Founding Father of the United States—was not guaranteed a good crop. In 1762, George Washington lamented his poor luck with tobacco: "I confess . . . it [tobacco cultivation] to be an Art beyond my skill, to succeed in making good Tobo. as I have used my utmost endeavors for that purpose this two or 3 years past; and am once again urged to express my surprise at finding that I do not partake of the best prices that are going."[26]

Frustrated with his attempts at a successful tobacco harvest, Washington dropped tobacco altogether in 1767, and turned to wheat. The future first president of the United States of America found this to be a more successful and lucrative crop, and one that allowed him free time to pursue his favorite pastime—foxhunting. Thomas Jefferson, another prominent Virginian planter and future president, agreed with Washington's preference for wheat over tobacco, but for different reasons: "It [tobacco] is a culture productive of infinite wretchedness. . . . Those employed by it are in a continual state of exertion beyond the power of nature to support. Little food of any kind is raised by them; so that the men and animals on these farms are badly fed, and the earth is rapidly impoverished."[27] For Jefferson, the possibility of a profit from the sale of tobacco was not worth the overemphasis on the plant's cultivation to the demise of the farmer who grew it. Enough planters in Virginia and Cuba were willing to make that potential sacrifice, however. Like many others hoping to make a profit from tobacco, these planters encountered many difficulties, from those inherent to tobacco cultivation to natural threats that were beyond their control.

Forces of nature played a significant role on the region's tobacco industry, particularly when they deviated from the norm. Oriente (the eastern district of Cuba) typically received forty to fifty inches of rainfall per year, while the western part of Filipinas and the areas south of Havana received the island's highest amounts, around sixty or seventy inches.[28] If it fell at the right time, rain was a boon for the region's farmers; if it did not, it could spell disaster to the year's crop. A lack of rainfall during the previous planting season meant that many *vegas* (tobacco farms) in Oriente remained unplanted in March 1752 at a time when the first set of seedlings should have been transplanted into the field and perhaps undergone topping. As a result, the region produced just over 93,000 *manojos* (bundles) of tobacco that season, the governor of Santiago de Cuba reported to Cuba's captain general. This figure was down from a typical year, which would have seen 150,000 bundles delivered to the royal monopoly. Frequent droughts plagued Cuba. A scarcity of rain in 1769 and again in 1792 resulted in a smaller than expected tobacco crop, disappointing not only the growers but royal officials as well. Officials in Bayamo in 1801 noted that the lack of rain during a significant period had set back the production and would result in a lower quality crop.[29]

Whereas the lack of rain resulted in a smaller crop, flooding sometimes associated with hurricanes or other storms also damaged the island's tobacco industry. Summertime deluges were particularly devastating because they occurred at a critical point in tobacco cultivation. Depending on when they happened, the floods could affect either the cutting of the first crop of tobacco in September and October or the transplanting of the second tobacco crop into the fields in June and July. Rivers in western Cuba, including the Cuyaguateje, Feo, and Herradura, frequently overran their banks and threatened nearby tobacco farms.[30] Early rains and subsequent flooding of the region's rivers left the surrounding *vegas* under "more than a *vara* [yard] of water," and negatively affected the 1775 harvest, new lieutenant governor Antonio Fernández noted in a letter he enclosed along with the annual *tazmia*.[31] Throughout the period under examination by this book, floods continued to plague Filipinas's *vegueros*. On 10 April 1779, between three and four in the afternoon, local rivers overran their banks, following three days of rain. The water covered the adjacent *vegas*, as well as the corn and rice plots, prompting one local official to state that although "it is necessary to conform to the will of God . . . these poor inhabitants do not need such misfortune since they are so unhappy, as it is much misery in which they live."[32]

Typical rainfall events could cause significant damage to the area's tobacco industry, and hurricanes, often frequent in number, brought even more destruction. Historically, most hurricanes affecting Cuba struck the western portion of the island. Havana contended with a 20 percent probability of a storm affecting the capital, while Filipinas had the highest risk of any area on the island at 38 percent. At least nine hurricanes struck western Cuba between 1763 and 1817. The storms occasionally came in close succession, bringing additional damage and flooding to areas already hard hit by earlier cyclones. Five consecutive days of "continuous rain" in mid-October 1774 caused the normally low Cuyaguateje River to rise to a "formidable level that has not been seen in many years," the *capitán del partido* (district captain or constable) wrote from Guane del Sur. A hurricane followed in short succession, bringing additional precipitation that began around nine o'clock on the night of 31 October, and continued throughout the following day.[33]

The weather's impact on the island and the western jurisdiction varied greatly, depending on each hurricane's track. For example, a wind- and rainstorm (possibly a hurricane) in November 1774 caused local rivers to overrun their banks, bringing devastation to the *partidos* (districts) of Guane del Sur and Guane del Norte (also known as Mantua). Farther to the northeast, however, officials in Consolación del Sur and San Juan y Martínez reported that luckily there were no losses in their locations.[34] Severe damage to an area's agricultural interests often preceded a cyclone's landfall. On 8 October 1846, as the San Francisco de Borja hurricane approached the island from the southeast before later passing over the capital, Havana's *Diario de la Marina* published a letter from a resident in Guane describing the damage already caused by the storm: "It is impossible for me to communicate the details concerning the lamentable effects of the rains of September. . . . It has rained without stopping a single moment for days and nights. The rivers have swollen to menacing proportions, the roads have become impassable. Tobacco farmers have lost almost everything," the resident indicated.[35]

An official in eastern Cuba alerted Havana during the summer of 1755 that the recent flooding of some of the rivers had damaged "many *vegas*."[36] At least four separate storms between 1768 and 1777 resulted in losses to the Cuban tobacco industry, both in terms of plants ruined in the field and damage to its infrastructure such as tobacco houses. A local official alerted Captain General Antonio María de Bucarely that a hurricane had hit Guanabacoa, located just east of Havana, on 15 October 1768. Twenty-eight tobacco houses were destroyed, while sixty remained standing.[37] This storm was soon followed by severe flooding to "levels we have never seen before" near Güines, which Francisco Gama lamented was a "second serious blow," that damaged the area's tobacco crop.[38] A hurricane in November 1774 brought destruction to Managua, leaving only two tobacco houses standing.[39] That same storm also damaged *vegas* in Rio Blanco, causing a local official to lament that the possibility of a good harvest was now lost. Tobacco plants in the field were damaged, as were the houses where the crop was stored in Govea and Güines. In Güines alone, 240 tobacco houses were affected negatively, as were 273,300 tobacco plants.[40]

In eastern Cuba, Mateo de Echavarría y Elguerra, the *factor* (local monopoly representative), informed Captain General Diego José Navarro of "a strong storm [*temporal*] of rain and wind that lasted seven continuous days . . . without weakening the drive of both elements." This storm seriously harmed the region's tobacco crop in November 1777. The local rivers had risen so much that the seedbeds and fields where tobacco was under preparation for planting were now "laid flat," and as a result, the following year's crop would be "reduced or none," de Echavarría warned.[41] During the 1790s, seven hurricanes affected the island and negatively influenced its tobacco industry.[42] Even typical storms posed a threat to the success of a year's crop; a letter written in April 1795 cites the harm caused to the tobacco seedbeds because of recent storms.[43]

Floods likewise brought devastating effects to the Virginia area's tobacco industry. A particularly destructive flood in May 1771 swept through the central and eastern part of the state. Most were unprepared for the devastation, believed to be caused by "heavy and prolonged rain in the Appalachian Mountains" to the west. The devastation was unexpected since the eastern part of the colony had been experiencing "serene skies." By late May, the James River had risen rapidly, with some accounts claiming as much as an increase of five feet per hour, while others described it as sixteen inches per hour.[44] Richmond's low-lying Shockoe Valley, home to numerous warehouses, was flooded. An article in the *Connecticut Journal* noted that Richmond's destroyed warehouses had held in excess of 1,600 hogsheads (800,000 pounds) of tobacco, and that "houses, warehouses, wine, tobacco, rafts of trees and timber, are constantly seen floating down the river." Several days later, accounts continued to trickle in about how severe this event really was. "From the mountains, to the falls, the low grounds have been swept of almost everything valuable; and the soil is so much injured that it is thought not to be of half its former value, & a great part is entirely ruined," the *Providence Gazette* told its readers. The flood destroyed or damaged in excess of 2.3 million pounds of tobacco along the James River alone. When the waters receded, the damage estimates added up to £2,000,000 (nearly $325,000,000 today) in direct losses, along with 150 fatalities.[45]

If climatic factors were not enough, tobacco was also subject to damage by pests. As one nineteenth-century writer noted, tobacco worms were "the only animal, save man, which is life-proof against the deadly nicotine of this cultivated poison."[46] One pest in particular, the green hornworm was especially destructive. William Tatham in 1800 described the process of eliminating hornworms from a tobacco field: "The act of destroying these worms is termed worming the tobacco, which is a very nauseous occupation, and takes up much labor. It is performed by picking everything of this kind off the respective leaves with the hand, and destroying it with the foot."[47] Another method of pest control employed by Virginia and North Carolina farmers was to let guinea fowl or turkeys into the fields to eat the worms.[48]

After months of cultivating the tender tobacco plants, the farmer reached another critical point in their transition to the final product—harvesting. Knowing when tobacco was at the appropriate state of maturity was difficult at best. As a nineteenth-century Virginia tobacco manual acknowledged, "it is difficult to convey an idea of ripe tobacco by description. It can only be learnt by observation and experience."[49] Farmers relied on several indicators to determine that the tobacco was ripe, although these signs did not occur simultaneously. Farmers from Cuba to Virginia agreed that a change in the color of tobacco leaves from gray-green to a yellowish-orange, sometimes in spots, was a sign that that indicated the tobacco was ready for cutting. So was the downward turning of the leaves toward the ground.[50] If the leaves had insufficient moisture when they were cut, they would be brittle and break, but if they were too moist, the tobacco would rot in the curing barn. In short, the farmer needed to balance his desire to place his tobacco within the safety of the barn against keeping it longer in the field so that it could reach the maximum point of maturity, when it would weigh more and therefore bring him a higher profit. The proper timing of harvesting tobacco was extremely vital, so much so that eighteenth-century Virginia planter Landon Carter and his contemporaries believed that a farmer's success—and, ultimately, his reputation—hinged on making the right decision regarding the proper time to cut tobacco.[51]

Cutting, as with other operations, differed in some respects from region to region. Period instructions to tobacco farmers from Cuba to Virginia recommended cutting during the day, although they differed on what was the appropriate time. Cuban farmers removed tobacco leaves in September and October from 9 a.m. until 2 p.m., while Virginia farmers cut tobacco either before ten in the morning or during the evening if the weather was hot but dry. Greater variation, however, was found in the method of removing the plant from the field. A Cuban farmer broke the leaves off the tobacco plant and held them on his left arm until hc had between forty-five and fifty leaves. He then laid the leaves on the ground to dry in the sun for six to eight minutes. The leaves were then placed on a piece of leather called a *burro* and carried to the tobacco house.[52] A 1788 document noted that a Cuban slave could carry between eight and ten *arrobas* (200 to 250 pounds) of tobacco on his head.[53] Virginia's farmers went between two rows, cutting the tobacco stalk to within six inches of the ground and then slicing the stalk upward to just below the bottom leaf. Workers stood the cut tobacco plant upright in the row or propped and turned it so that the sun could make the leaves pliable for handling. When ready, the stalks were placed in a circle, and covered with green brush to prevent sun scorch.[54]

Both Cuban and Virginian growers took their product to four-sided wooden barns, where they placed the tobacco on sticks stretched between poles that ran parallel to the ground. During the late eighteenth century and early nineteenth century, tobacco curing used one of two methods. Air curing was practiced in both Cuba and Virginia. This method allowed air to circulate through the barn and the hanging leaves, but not during the night or in colder temperatures. It was important that the tobacco be sufficiently dry and adequate space given between each pole to provide ample ventilation. If not, mold could develop on the stems and leaves, which would ruin the quality of the product. The other method, used throughout the United States, was fire curing. The barn was made airtight, the sticks of tobacco packed tightly together, and heat introduced into the barn, with the temperature raised slowly over time until the leaves turned a light brown.[55]

The beginning of Cuba's rainy season during May and June spelled an increase in atmospheric humidity, making the now-dry leaves more flexible, allowing for safe movement of the crop without risking breakage. As the farmers began to prepare their product for transport and sale to the monopoly, they carefully separated the tobacco leaves into different piles based on their quality and covered them with plantain leaves to protect them. If the tobacco was in good condition, the farmer used dry plantain leaves, but if the tobacco had become too dry and needed additional moisture, he used green plantain leaves. The piles were placed three fingers high on elevated benches, where they were wet down with a brush or sponge on the top and bottom. The tobacco was then returned to the tobacco house to hang. After the curing process was complete, and the tobacco ready for market, *vegueros* placed it into manageable bundles. Each leaf was lined up with the others at its widest part, with the central veins together. Depending on the quality, between twenty-five and forty leaves were grouped together in this manner, and tied at the top with an additional leaf to form what was known as a *gavilla* (sheaf). Four *gavillas* tied together with string were known as a *manojo* (bundle).[56]

Cuba's *vegueros* presented their tobacco to the monopoly at purchasing fairs held between August and the end of October, providing the tobacco survived the numerous threats ranging from a tobacco farmer's poor decision-making to problems with rain and pests.[57] The money to pay the local *vegueros* for their crop was ideally already in place, delivered from Havana by trusted men escorted by armed cavalrymen. Such was the case during the 1750s when *arriero* (mule pack driver) Pedro Hernández and Lieutenant (junior grade) don Manuel de Quintana regularly delivered funds to eastern Cuba to pay for the area's tobacco harvest (see Chapter 4). José Aguilar and Fabian Herrera similarly arrived in Puerto Príncipe in August 1769, carrying money designated for the purchase of local tobacco. From there, the pair headed east toward Santiago de Cuba on 5 August, escorted by four dragoons.[58]

After tobacco left the hands of the *vegueros*, it made its way to the royal tobacco factory in Havana. Horses or mules and their drivers transported tobacco from the rural areas around the city to the capital. For example, in mid-November 1775, the captain general alerted *alcaldes* (magistrates) in San Felipe and Santiago de Bejucal of the need for *arrieros* to bring tobacco from the Vuelta Abajo to the Real Factoría (Royal Factory). In less than a week, a local official, Domingo de la Cruz, responded that the needed *arrieros* and their horses had been secured. A total of sixteen men and one hundred and twenty-five horses led by *arriero* foreman Juan González left nearby Santiago de Bejucal on 29 November and headed for the Vuelta Abajo.[59]

Not all attempts to secure *arrieros* to transport tobacco into Havana were as simple. Even across short distances, transportation problems often prevailed. One man noted that it was "very difficult" to bring tobacco from the district of Macuriges to Havana, some forty leagues distant. The main problem, the local official stressed, was an ineffectual transportation system, which he deemed "impossible."[60] In the spring of 1776, it was clear that there were problems. The Marqués de la Torre assured San Felipe y Santiago's *alcaldes* that mule pack drivers would come through Güines with approximately twenty-three animals to collect green tobacco, while officials in Bejucal were notified to expect *arrieros* with about fifty animals. A local official in Bejucal complained on 26 April that several *arrieros* were not available, and others did not have the animals needed to carry tobacco to the city.[61]

Issues such as damage to the island's roadway infrastructure repeatedly added to transportation difficulties. In late December 1757, for example, a Cuban official reported that the collection of tobacco from Güines had been suspended and that it would remain so until the roads were "transitable." In February 1801, the island's intendant, Luis de Viguri, informed Miguel Cayetano Soler, Spain's finance minister of the urgent need to repair the road from the Surgidero de Batabanó into Havana, made "impassable" by recent storms.[62]

Tobacco traveling longer distances, such as that from the eastern part of the island, came by sea. Throughout the summer, fall, and winter, workers placed tobacco gathered and purchased from eastern *vegueros* on ships leaving from Bayamo, Santiago de Cuba, Puerto Príncipe, or Holguín, and sent it west toward Havana. These vessels off-loaded their cargoes at the Surgidero de Batabanó on the southern coast. From there, tobacco went overland to the royal monopoly building in Havana. An example of this took place in June 1767, when vessels of varying sizes—ranging from a small *lancha* (launch) to a *goleta* (schooner)—carried tobacco from Bayamo down the winding Bayamo River, while three other ships stood ready to take additional cargoes. Several years later, Captain don Joseph Garcia guided the schooner *San Francisco de Paula* through the Bay of Santiago de Cuba. The ship sailed past the Castillo de San Pedro de la Roca (the city's multilevel fortification, perched high on a cliff overlooking the water) and into the open Caribbean Sea before turning west. On that day, 25 August 1770, the ship's hold contained a load of twenty-nine hundred *arrobas* (approximately 73,515 pounds) of tobacco destined for the royal monopoly. Additional ships made the same journey in January 1774, and again in August 1775. Tobacco coming from Puerto Príncipe and Holguín, on the other hand, frequently left in the fall.[63]

Once tobacco reached Havana, it was delivered to the Real Fábrica de Tabacos. The early physical headquarters of Cuba's royal tobacco monopoly were rather inauspicious. A "dilapidated building that served as a warehouse" for the monopoly was located near the royal shipyard, close to the southern end of the *Campo de Marte*, the military practice and parade ground that marked the boundary between the rural outskirts and the capital itself.[64] During the 1770s, Spain authorized construction of a new edifice to house the royal tobacco factory and monopoly headquarters. The chosen site was located in Havana's Jesús María neighborhood, outside of the city's *muralla* (protective walls). The property, which had Factoría, Revillagigedo, and Diario Streets among its borders, was situated south of the arsenal and near the *Puerta de la Tierra*, one of the entrances into the fortified capital. The colonial government seized and razed at least eight houses in March 1773 to make way for the structure, which would cover an entire block. The facility was designed by director of engineers Silvestre de Abarca (the man responsible for Havana's refortification following its return from the British) and royal engineer Luis Huet.[65]

The newest headquarters of Cuba's royal tobacco monopoly had access to the sea via the Tallapiedra wharf. Nineteenth-century visitor Alexander von Humboldt described the vast building as one of the city's "principal edifices . . . less notable for [its] beauty than for the solidity of [its] construction." Other reports from the period seem to prove Humboldt accurate, at least in terms of the building's imposing size and strength. A large structure with massive, high walls, the factory was accessed by way of a road that ran toward the *muralla* near where officials had proposed a new city entrance. A large wooden water-powered crane on the nearby *Muelle de la Factoría* (Monopoly Wharf) unloaded tobacco barrels from the interior that arrived by sea. All tobacco, whether it was destined for Spain or elsewhere, passed through the property's well-secured entrances into the lower level of the factory building. This area contained several large, open interior squares, including a central one complete with a fountain. Water-powered *molinos* (mills) in Cuba's interior had pre-ground some of the island's tobacco; at Havana's factory, horse-driven stone mills on the building's ground level reground the tobacco to a finer consistency. The ground level boasted twenty-foot ceilings and brick floors, while free and slave workers produced cigars on an upper story.[66]

The monopoly considered tobacco safe once it reached the Havana headquarters of the monopoly. Yet, the historical record demonstrates that tobacco was not shielded from the ravages of nature even in the possession of the Real Factoría. A severe storm or unexpected flooding could easily wipe out a sizable portion of the tobacco stored inside the monopoly's buildings. Between 15 and 21 September 1778, rains "like never seen before" virtually destroyed the contents of the new tobacco monopoly building, constructed just several years before, and four warehouses, all of which were full of the best-quality tobacco. The "weight of the water undermined the walls" of warehouses one through four, while in *almacenes* (warehouses) five through nine, and in warehouse thirteen, there was no way for the water entering them to escape. In the warehouses without roofs, water rose to a height of two *varas* (five and a half feet). Officials ordered the *zanja* (ditch) opened, and the remaining tobacco transported to the headquarters of the Real Hacienda (Royal Treasury) since the monopoly's own facilities were full, but it was with little effect. The damage was done, but monopoly officials quickly responded. Captain General Diego José Navarro was reassured just three days later that the buildings were now dry and again ready for tobacco.[67]

Once Havana's Real Factoría processed the island's tobacco, workers loaded it onto ships, preparing it for the long voyage to the empire's main tobacco monopoly headquarters in Spain. From Havana, the ships headed through the Straits of Florida and out into the open Atlantic. Their destination was Cádiz, the seat of all commercial activity with the peninsula since 8 May 1717, when King Felipe V moved two key imperial institutions, the Casa de la Contratación (House of Trade) and the Consulado (merchant guild), from Seville to the southern Spanish port city. Tobacco then traveled north to the city of Seville. There, workers processed tobacco in the empire's main *estanco* headquarters. This location, the enormous Real Factoría de Tabacos, was a building described by a particularly critical nineteenth-century traveler as "utterly devoid of architectural merit."[68]

The process of growing, curing, shipping, and processing tobacco was a complicated one, risky at every step. These methods, however, remained in essence unchanged from those utilized as early as the sixteenth and seventeenth centuries. In areas as disparate as Cuba and central Virginia, similar eighteenth- and nineteenth-century agricultural practices demonstrate the continuity in tobacco growing and cultivation over time. A tobacco farmer's knowledge rested on experience, patience, and skill, and was passed from one generation to the next. Europe's colonial empires existed within relatively static geographic boundaries. Yet, the knowledge needed to secure a successful tobacco crop flowed between the north and south in the early colonial Atlantic world, linking such competitors as Cuba and Virginia and the crop they cultivated. Economic entities like tobacco monopolies sought to regulate such vital endeavors, but it was not possible to subject intellectual information to imperially imposed confines. When Spain sought to extend the successful cultivation of tobacco into new areas, its officials sent Cuban growers to Peru to instruct farmers there how to cultivate tobacco. Ultimately, the knowledge possessed by Cuban *vegueros* and the particular agricultural techniques and manufacturing processes they employed were the elements that made a Cuban tobacco product distinctly Cuban.

3 "One of the good growers"

FOR MANY ON THE ISLAND OF CUBA, in Spain, and beyond, tobacco cultivation was a way of life. The growing and manufacture of tobacco products encompassed a large portion of Cuba's colonial society, drawing together a disparate group of people across nearly all social, racial, and economic sectors. Despite their disparate backgrounds, tobacco unified them. This view of Cuba's tobacco farmers runs counter to the existing historiography best exemplified by Fernando Ortiz's seminal work, *Contrapunteo cubano (Cuban Counterpoint)*, which described the island's *vegueros* as predominantly poor, white Canary Islanders. Ortiz and many other historians since have repeated the claim that tobacco farmers were small-scale farmers, growing limited amounts of tobacco. They were so poor, the historiography contends, that they were barely able to sustain their families, much less afford to buy slaves to work in the fields. Spanish and Cuban archives, however, hold documents that present a different picture of the *vegueros*, and reveal that tobacco cultivation was not just a "pursuit of the lower orders."[1]

Tazmias were used by the monopoly to record tobacco production. When they are combined with period censuses, they provide a method of determining the composition of those owning or renting land on which tobacco was grown. While *tazmias* were designed to record information about tobacco production rather than those who cultivated it, researchers can learn much from these documents about the race, social status, gender, and occupation of these agricultural producers. Occasionally, clues to the national or regional origins of these farmers are provided through period descriptions such as *el isleño*, (the Canary Islander), *el español* (the Spaniard), *el portugués* (the Portuguese), *el gallego* (the Galician), and even *el inglés* (the Englishman).[2] The numerous annual *tazmias* available in Spanish and Cuban archives demonstrate conclusively the presence of people of color, both free and slave, in all parts of Cuba. Correspondence similarly confirms that from the 1750s onward, residents and government officials alike sought slaves for use in the tobacco fields. Free and enslaved Africans worked in tobacco fields across the island, even in Filipinas.

The diversity of Cuba's tobacco farmers meant that tobacco regulations affected a wide swath of Cuba's colonial society and population. Through an examination of these important colonial actors, the Bourbon reforms in general and Cuba's tobacco monopoly in particular can be viewed in historical context. In this light, the Cuban people no longer appear as helpless bystanders who were forced into compliance with the Bourbon reforms because no other option was available to them.

Tobacco farmers and the land they worked constituted an important share of the rural colonial Cuban countryside. The island's 1778 census included an unknown number of tobacco farms within the 3,958 *estancias* (cattle ranches) reported. The 1792 census ordered by Captain General Luis de las Casas just over a decade later noted a combined total of 6,134 cattle and tobacco farms. While it is difficult to specify exactly by how much the number of *vegas* grew during this period, the overall number of Cuba's *estancias* increased by nearly 55 percent in a mere fourteen years. By the early nineteenth century, tobacco farms constituted their own separate category on the island's censuses, and in 1811, Cuba's 5,534 *vegas* covered nearly 371,560 *caballerias* (approximately 12.4 million acres), or 44 percent of the island's total land area. The number of farms dedicated to the plant decreased by early January 1812, as the end of the monopoly loomed near; a count of those involved in the island's tobacco industry reported 3,996 *vegas* and 8,775 people engaged in tobacco cultivation. The majority of these tobacco farms, 80 percent, were located in the central and eastern parts of the island. Local officials were keenly aware of the decrease in production, and they opined that if additional lands were made available for tobacco cultivation, a total of twenty thousand *vegas* tended by forty thousand farmers was possible.[3]

The tobacco *vega* remained an important rural fixture in late colonial Cuba, not only for the crop it produced, but also as the home of thousands of residents. By 1846, approximately 17 percent of Cuba's combined rural community (71,319 people) lived on *vegas*. Even in distant Oriente (East), an area not usually associated with tobacco cultivation, more than 20 percent of the region's total rural population resided on a tobacco farm. Oriente's jurisdictions of Guisa, Sagua, and Mayarí Abajo were predominated by tobacco, where more than 75 percent of rural residents lived on a *vega*. When the next census was taken in 1862, over 108,000 residents—including 75,000 rural whites and 17,675 slaves—lived on tobacco farms throughout the island, and nearly one-fifth of eastern Cuba's free black population lived on *vegas*.[4]

Several areas of the island were centers of tobacco cultivation. One of those, Filipinas (also known as Pinar del Río), is discussed in greater detail below in the examination of the interconnected nature of tobacco and the local community. However, another area associated with tobacco was the *villa* (small town) of San Julían de los Güines, long described as the epicenter of eighteenth-century sugar expansion. Yet, in just four years—1768 to 1771—the number of individuals listed on the *tazmias* as tobacco farmers in Güines increased approximately 52 percent, from 133 to 202. The overall number of *vegas* dropped slightly from 279 to 248 between 1771 and 1774, yet more workers labored in the area's fields—682, up from 586 in 1771. Tobacco remained an important agricultural pursuit for Güines despite this slight decrease in the number of farms, and by 1775, 87 percent of the town's residents lived on *vegas*, which occupied one-fifth of the *villa*'s total land area. Tobacco's role in Güines only grew stronger when the area was reorganized in 1786 in an effort to focus specifically on tobacco. The plan was a success, and by late June 1792, the number of tobacco farms in Güines grew to 590, and exceeded 600 only four years later. As late as 1809, tobacco remained an agricultural presence in Güines, with sixty farms still raising the crop.[5]

Who were these people who cultivated and sponsored the growth of tobacco in colonial Cuba's agriculture and economy? Ortiz's poetic description of the Cuban tobacco farmer is in large part responsible for the creation of one of the most lasting images of the *veguero*. His depiction reveals his own sense of pride in these farmers and a romanticized longing for a past that Ortiz believed was slowly disappearing. Throughout *Cuban Counterpoint*, the tobacco farmer was contrasted with the sugar planter:

> The tobacco-grower was a simple countryman who required no machinery beyond a few tools and who could supply his own needs from the limited resources of the local general store. Whereas the sugar-planter acquired wealth, titles of nobility, government posts, refinement, and, at times, a desire for progress, the *veguero* was always a small, rustic, rule-of-thumb farmer. While the planter gave Cuba railroads before they were introduced to Spain, and Havana had its flourishing theater presenting plays and operas as good as those of Madrid, the *veguero* still rode his horse through the woods and found his entertainment in cock-fights, songs, and country dancing.[6]

It is this image of the *veguero* as a poor, white, small-scale farmer that is repeated throughout most of the existing scholarship.[7] Yet social, economic, and racial diversity was the rule among those who grew and sponsored the production of Cuban tobacco, and indeed, it was this lack of homogeneity among those in the island's tobacco industry that allowed Spain to exert social and economic control over an even larger group of people. Had those associated with tobacco been limited to just a few types within Cuban society, the reach of the monopoly would not have been nearly as wide-ranging.

A location closely associated with tobacco that illustrates the close connections between tobacco cultivation and all sectors of the local community was the area now known as Pinar del Río. Tobacco ultimately provided the very backbone that shaped and maintained this rural society located on the far western end of the island. The plant tied those from very different social, ethnic, racial, and economic backgrounds to one another. Through the cultivation and promotion of tobacco, *vegueros*—free and slave, white and black, rich and poor—competed with one another to bring in the largest harvest and profit for their hard work. An examination of the sources that record this tobacco production demonstrate the interconnectedness of this region's agricultural producers and community at large. In such an isolated territory, few other economic pursuits could have led such a diverse group of men as San Juan y Martínez residents *isleño* Domingo González, Lieutenant Manuel del Castillo, free black Diego de Mesa, and Father José Cipriano de Ignocencio to compete economically during the 1770s and 1780s.

Situated fifty-one leagues west of Havana, San Juan y Martínez was approximately halfway between Mantua (the most distant *partido* of the new jurisdiction) and Consolación del Sur and Consolación del Norte (both just forty leagues from the capital). The district was located in the shadow of the Sierra de Guane and Alturas de Pizarras del Sur Mountains. These fertile lands, fed by the San Juan y Martínez, San Sebastian, and Seco Rivers, produced what one traveler touted as "the best tobacco in the world." The town of San Juan y Martínez was one of the region's smallest; it was virtually nonexistent, with merely one lone house and a church even as late as 1789, while the largest *villa* in the jurisdiction, nearby Pinar del Río, boasted twenty houses. The surrounding countryside, however, supported a vibrant agricultural community of between 426 and 462 souls who lived on the many tobacco *vegas*, approximately a dozen *hatos* (cattle farms), and several *corrales* (small estates). As was the trend elsewhere in the region, the number of *vegas* in San Juan y Martínez dropped from the 1770s through the end of the 1780s. There were thirty-two *vegas* in 1770; the total dipped to a low of twenty-two in 1779, but managed to recover by the end of the decade, increasing to around fifty. San Juan y Martínez's tobacco *vegas* remained an important nucleus of life for its rural residents into the mid-nineteenth century, with 97.4 percent of the *partido*'s inhabitants in 1846 residing on the area's 560 tobacco farms.[8]

The 1770 tobacco crop of 12,090 *cujes* (sticks) proved one of the largest of the decade for San Juan y Martínez, although other districts in Cuba's far western

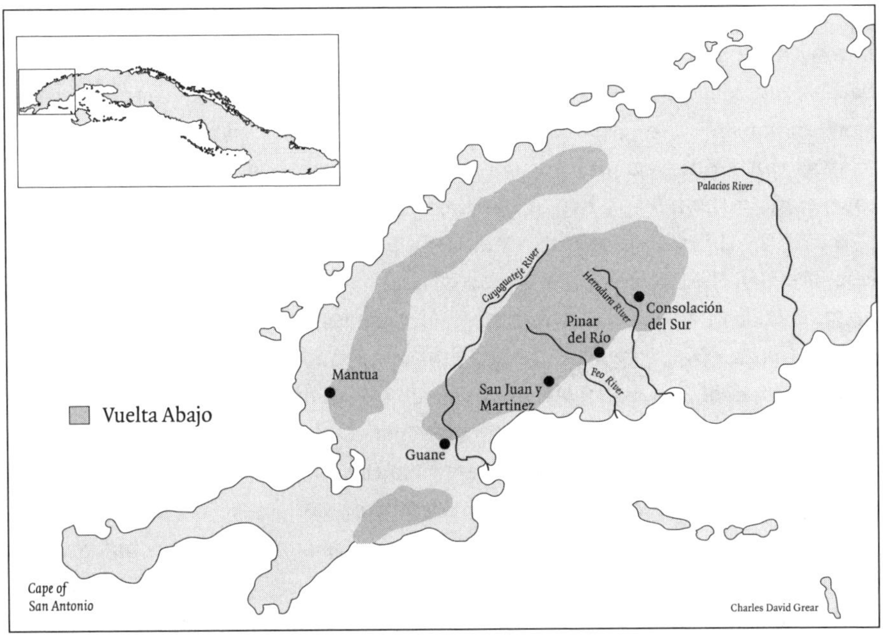

Pinar del Río. Map by Charles David Grear

region produced more tobacco. That year, the majority of the district's *vegueros* (twenty-three out of thirty-seven named individuals) did not work alone on their tobacco land, using between one and three additional laborers. Eight *vegueros* utilized slave labor, primarily hired from fellow tobacco farmers and slave owners, while two tobacco farmers used free laborers to assist them. Four pairs of men were partners, each co-owning the *vega* they worked together, whereas five tobacco farmers relied on family members—primarily sons, although one used several nephews—to supply needed hands in the field.[9]

One particular tobacco farmer, Domingo González, provides a clear example of the diversity found among the area's *vegueros* and the various types of labor employed in their fields. An *isleño*, slave owner, employer of hired slave labor, and militia member, González was born around 1731 or 1732 in the Canary Islands. He immigrated to Cuba by the mid-1750s, where he spent at least two decades making a livelihood from growing tobacco in the San Juan y Martínez district. By the time he was in his mid-twenties, he had accumulated capital in the form of 426 *pesos* in silver, and he married María Asuncíon de los Reyes in 1757. González would eventually own at least five slaves, and a tobacco farm in the *partido* of San Juan y Martínez valued at three thousand *pesos*.[10]

By 1770, González began to experiment with slave labor on his farm, located near the San Sebastian River. That year, he hired two slaves from fellow tobacco farmer Captain Juan de Mesa, who supplied many of the area's hired slaves. The project proved successful. González yielded eight hundred *cujes* of tobacco, tying him with Lieutenant Manuel del Castillo for the title of the district's largest tobacco producer. Combined, González and del Castillo, assisted by four slaves, were responsible for 13 percent of the *partido*'s harvest. Their nearest competitor was thirty-one-year-old Antonio Rubi, who produced six hundred sticks of tobacco with the help of two slaves hired from Captain Nicolás Hidalgo Gato. Growing the third-highest amount of tobacco, five hundred *cujes* each, were four *vegueros*—don Nicolás de Otaño, Simon de Acosta, Nicolás Iglesias, and Juan Quintana. Only de Otaño used slave labor; Quintana turned to two sons, and Acosta and Iglesias co-farmed with Antonio Pérez and Francisco Fletes, respectively.[11]

The district of San Juan y Martínez was unable to meet the previous year's success in 1771, but its production was more evenly distributed among the thirty-one *vegueros*. González grew slightly less than the year before, but his seven hundred sticks of tobacco was nearly 10 percent of the district's entire crop that year, making him the largest tobacco producer. He employed three unidentified workers—perhaps the same slaves he had hired the year before—on his *vega*. In 1770, free black Diego de Mesa had been one of seven *vegueros* competing for the title of fourth-largest tobacco grower; in 1771, the *partido*'s drop in overall production allowed him to tie with Lieutenant Manuel del Castillo as the *partido*'s second-largest producer. The previous year, Mesa had used a slave hired from a local owner, and it is possible that the unnamed additional worker in 1771 likewise was a hired slave.[12]

Throughout the 1770s and 1780s, Domingo González continued as one of the largest tobacco producers in the San Juan y Martínez district. González easily led his fellow *vegueros* by 1779, growing four hundred *cujes* of tobacco for the royal monopoly. His closest competitor, Joaquín Gutiérrez, yielded 350 *cujes*, while longtime rivals Nicolás Iglesias and Antonio Rubi contributed just 200 and 50 *cujes*, respectively. The following year, 1780, Iglesias's crop grew to a remarkable eight hundred *cujes*, surpassing González as the largest producer. Joaquín Gutiérrez no longer grew tobacco, and Rubi held steady at a mere fifty *cujes*, among the district's smallest crops for that year. The entrance of local constable Vicente Blanco into the area's tobacco industry in 1781 and 1782 challenged González's reign as the district's leading tobacco producer. Blanco's *vega* managed to produce six hundred *cujes* on land worked by just three people, while the four people working on González's farm yielded five hundred *cujes*, making Blanco the leading tobacco grower in 1781 and pushing González into second place. The following year, 1782, Blanco and another farmer, fifty-year-old Canary Islander and slave owner José Oropeza, each turned in five hundred *cujes* of tobacco. González held on to second place with four hundred *cujes*.

Longtime tobacco farmer Nicolás Iglesias's workforce declined from a high of three workers, from the late 1770s through 1781, to a low of just one person, in 1783. He had been one of the top tobacco growers in the previous decade, but in 1781, he tied for sixth place (two hundred *cujes*), a position he shared with four other farmers. The following year, his crop yields fell even more and he dropped to eighth place, sharing it with five other *vegueros* who produced just one hundred sticks of tobacco. In 1783, his production numbers improved, with Iglesias and González each growing three hundred *cujes* of tobacco.[13]

Tobacco cultivation was a family affair for the Domingo González household, with his wife, María Asuncíon de los Reyes, owning *vegas* in her own right. Born in the San Juan y Martínez district, de los Reyes was the legitimate daughter of Francisco de los Reyes and Teresa de Mena, both natives of the city of Havana. By late January 1777, she had dictated her will before a local notary. She requested burial in the habit of San Domingo in the *iglesia auxiliar* (auxiliary church) of San Juan Bautista, along with three masses of the soul. She had brought no capital of her own to the union with González twenty years before, but her possessions now included two unbaptized slaves named María del Rosario and José Narcisco, three mares, two riding horses, a two-year-old colt, and two *vegas*. One of the *vegas* had a tobacco barn described as being in poor condition, while the other included a house, a tobacco barn, and a separate kitchen building.[14] She updated her will on 19 October 1785. Two more sons, Pedro and Domingo, had been born in the years since, and this time she requested burial in the chapel of Nuestra Señora de la Candalaria in the Church of San Juan. Her financial holdings had increased significantly; among her real estate assets were two rural properties she may have inherited from her now-deceased parents. One was a *vega* that contained a house and two tobacco barns of varying condition. The other was described as a farm of three *caballerías* (approximately one hundred acres) located in the jurisdiction of Santiago and bordering land belonging to a marquis. It had been purchased at public auction for 2,840 *pesos* and contained two houses, a kitchen, and a tobacco barn. Additionally, de los Reyes owned a house in the town of San Juan, two male slaves named José and Carlos, and seven yokes of oxen.[15]

Other members of the local community also participated in the area's tobacco industry, including the district's only priest, Father José Cipriano de Ignocencio. The cleric had lived in San Juan y Martínez since at least 1775. Sometime between 1781 and the spring of 1782, Cipriano de Ignocencio purchased a *vega* in the *partido* of San Juan y Martínez, and two unidentified people labored in his fields. His participation in the area's tobacco industry was short-lived, as 1783's *tazmia* indicated that he was no longer producing tobacco for the monopoly. The short period of ownership was not uncommon. Period records indicate that the composition of those owning colonial Cuban tobacco farms was extremely fluid, as individuals bought and sold *vegas* from one year to the next. Father Cipriano therefore was no exception. Many of the other top producers in previous years—men like Joaquín Gutiérrez, Antonio Rubi, and Captain Vicente Blanco—likewise were no longer engaged in the cultivation of tobacco in San Juan y Martínez by 1783. Nicolás Iglesias and Domingo González, however, remained constant fixtures in the district's tobacco fields, continuing to produce tobacco for the monopoly. The diversity among those sponsoring and cultivating tobacco demonstrated how entrenched tobacco was in the local community and the ways in which the plant linked its disparate residents.[16]

In many parts of western Cuba, tobacco likewise dominated the local community and its activities. Another town with strong roots in the tobacco industry was Pinar del Río. Located along the winding Guamá River, Pinar del Río was about forty-five leagues west of Havana, in the eastern foothills of the Alturas de Pizarras del Sur. Frequent flooding from the nearby river prompted its move to a higher location in 1726, and again in 1746, when it reached its final site.[17] The settlement was modest, and as late as the end of the eighteenth century, Pinar del Río was described as "a small town that consists of 20 to 25 houses and its parish church, all made from palm trees."[18] Tobacco provided an important source of income for its residents as the area's primary agricultural pursuit. By 1770, in one district alone, San Rosendo de Pinar del Río, sixty-eight tobacco farms employed a total of ninety-seven people and produced three thousand *arrobas* (76,050 pounds) of tobacco. The district supported the greatest number of tobacco farms, seventy, of any in the jurisdiction by 1787, surpassing second-place San Juan y Martínez by twenty-one *vegas*. The following year, Pinar del Río's lead grew even more to eighty-four *vegas* in all, followed by Consolación del Sur with fifty-eight, and San Juan y Martínez at forty-eight.[19]

The shared pursuit of tobacco cultivation connected the residents to one another, as it had in San Juan y Martínez. Tobacco linked former slaves like María Loreto with slave owners and *vegueros* such as Jacinto Morejon, Anzieto Morejon, and Agustín Murquia, as well as with free blacks such as José Antonio Ribero. As in San Juan y Martínez, females were among those responsible for providing tobacco to the monopoly. *Parda* (mulatto) María Loreto, like María Asuncíon de los Reyes, was the wife of a *veguero* as well as a participant in the area's tobacco industry in her own right. The details of Loreto's life reveal the intertwined relationships between enslaved and free black members of rural Cuban society and their white counterparts, along with a definitive connection between tobacco and freedom. Jacinto Morejon, one of the area's *vegueros*, purchased her when she was approximately seventeen years old, around 1775. Morejon was a long-standing tobacco farmer, having grown tobacco in the *partido* of Consolación since at least 1770. Period records hint at the likelihood that some of the funds used for Loreto's 1788 manumission came from economic transactions related to tobacco. In 1779, the still-enslaved Loreto grew tobacco on her own account, making her one of only two *vegueros* of color and the only black woman among the thirty tobacco farmers in the Consolación district. This decision to enter the area's tobacco industry showed resourcefulness, as she took advantage of her proximity to tobacco and her owner's choice of profession to begin to earn the money necessary to purchase her freedom. On 18 July 1788, Loreto paid Jacinto Morejon three hundred *pesos*, and succeeded in gaining emancipation. The now-thirty-year-old woman was free.[20]

Loreto's ties to the area's tobacco industry also extended to her husband, *moreno libre* (free black) *veguero* José Antonio Ribero. In 1784, Ribero entered the area's tobacco industry as a tobacco land owner, acquiring a *vega* from *veguero* Dionisio Manresa. The property was southeast of the *villa* of Consolación del Sur and bordered the Herradura River and the royal road. Described as enough tobacco land for one man to work, it was sold by Ribero on 27 March 1786 to local tobacco farmer don Anzieto Morejon for 170 *pesos*. Ribero died between the spring of 1786 and fall of 1788, leaving Loreto widowed. She continued the pattern of tobacco land transactions initiated by her deceased husband in the coming years. Loreto sold additional tobacco land to Anzieto Morejon on 18 October 1788. This property was located in the same area as the *vega* Morejon had purchased two years earlier from Loreto's husband and was described as large enough for one man to work. Morejon bought the property for 160 *pesos*, but was unable to pay for the entire transaction. He signed a promissory note for ninety *pesos* that same day. Under the terms of the agreement, he agreed to pay Loreto twenty *pesos* at the end of eight days, twenty additional *pesos* in two months, and the remainder from funds Morejon would receive for his 1789 tobacco crop. Captain Bartolomé Diaz, a fellow *veguero* and slave owner, acted as María Loreto's representative in both 1788 transactions.[21] These *tazmias* and real estate records demonstrate the very real possibility that Loreto, her husband, or both earned the money to purchase her freedom through the sale of tobacco to the royal monopoly. Additionally, these documents allude to the status and integration of the Ribero-Loreto family into western Cuba's society, which came from their position as free people, property owners, and *vegueros*.

The distance between San Juan y Martínez and Pinar del Río did not impede economic transactions between the towns' residents. Agustín Murquia, like *veguero* Domingo González (described earlier), had spent many years working tobacco and owned a *vega* in the Pinar del Río district by 1770. In April 1775, González, who lived in San Juan y Martínez, purchased a Mina slave from Murquia.[22] The transaction transferred the slave, Antonio Bentura, from one of Pinar del Río's most productive *vegueros* to an even more prolific one in San Juan y Martínez. Bentura's new owner, González, filed manumission papers with the local notary even before the month was out. All that remained was for Bentura to procure the necessary funds to purchase his freedom. Over the next few months, the forty-year-old slave accumulated the needed 150 *pesos* to purchase his manumission, and on 17 October he became a free man.[23]

Throughout the Americas, slaves engaged in agricultural pursuits for their own benefit. Like María Loreto, Bentura probably earned the money to buy his freedom by choosing to grow tobacco for himself as well as his owners. Unlike other agricultural crops that were grown for personal consumption, tobacco offered Bentura a guaranteed market for his product via the monopoly and the certainty of a financial reward for his work.

TOBACCO DOMINATED WESTERN Cuba's agriculture well into the middle of the nineteenth century. In agricultural areas, there were more male than female slaves, with the men outnumbering the women by a ratio of over two to one in western Cuba during the 1770s. This gender disparity in the slave population continued well into the nineteenth century, with an 1846 ratio of forty-three females to one hundred males, and forty-seven females to one hundred males in 1862. Most of these slaves, particularly males like Bentura, probably labored on the region's *vegas*. The phenomenon of male slaves laboring in agriculture was not unique to Cuba, but repeated elsewhere in the Americas, where the majority of male slaves worked in agricultural production and were less likely to receive manumission than female slaves.[24]

Western Cuba offered just one example of the diversity among those who grew and sponsored the cultivation of tobacco. A detailed examination of this area reveals competition among tobacco producers across all levels of colonial society, as well as economic and social interactions brought about by their collective involvement with the area's tobacco industry. Additionally, for some of those held in bondage, tobacco growing quite possibly led to their ability to purchase their freedom. Many other locations across the island similarly exhibited these characteristics and patterns. All these findings challenge the current historiography, which simplified the depictions of those who labored in the island's tobacco fields, and those who promoted the cultivation of this valuable plant.

Cuba's colonial *vegueros* were not limited to those who were poor, white, and Canary Islanders, but a depiction of these valuable agriculturalists that did not include these groups likewise would be erroneous. Fernando Ortiz relied on a long historical connection between Canary Islanders and producers of tobacco. From the earliest days of Cuba's colonization, *isleños* made their mark on the island's population. Approximately a quarter of the immigrants entering Havana between 1585 and 1645 were Canary Islanders. *Isleño* immigration received a significant push when a 1681 royal *cédula* (bill) directed ship captains leaving for the West Indies from the Canary Islands to take five emigrating families for each hundred tons of cargo carried. The captains were exempt from customs duties for ten years in return, and by 1693, over eight hundred *isleño* families called the area around Havana home. Other methods used to encourage Canary Islanders to come to the

New World included free passage to the West Indies, an exemption from taxes, and a cash subsidy of fifty *pesos* per family. Among the Canary Islanders who came to Cuba, the opportunity to grow tobacco was a key attraction, and the *isleños*, long considered expert agriculturists, quickly became known as expert tobacco farmers. Furthering the close association between tobacco and *isleños*, the word commonly used to describe a tobacco farm, *vega*, originated from the Canary Islanders.[25]

The presence of *isleños* in Cuba's tobacco fields, however, had diminished considerably by the late eighteenth century. Documents and *tazmias* for the period under study list few Canary Islander tobacco farmers. In the more than sixty *tazmias* consulted for this book, only thirteen individuals were described as *isleños*, out of the hundreds of *vegueros* listed. All the parts of Cuba were represented by these Canary Islanders, although many of them resided around Havana or in the central part of the island. Seven of the *isleño* tobacco growers lived in the areas surrounding Trinidad, Santa Clara, and Puerto Príncipe, while four lived around Govea or Cano, just outside of Havana, and only one individual resided in far western Cuba, the area most closely associated with tobacco. Nearly all the *isleño vegueros*—the eleven men for whom the data is available—worked alone or with one additional person on their tobacco farms, and in only two instances did an *isleño*-run *vega* employ between three and five workers (figures that included the owner).[26]

Among the *isleño vegueros* growing tobacco in central Cuba between 1777 and 1783, production varied greatly. Francisco el *isleño* tended just two thousand plants in the Partido de Samaraguan, near Puerto Príncipe. Close to Trinidad, Antonio's Rio de Caracusey-area *vega* contained nine thousand plants, and Bartholo cared for ten thousand tobacco plants near the Rio de Animado. Larger crops were possible, if additional workers were available. Salvador and Manuel, Canary Islanders who owned *vegas* near Puerto Príncipe in 1770, validate this. Manuel's *vega* was worked by two people, most likely including Manuel himself, and contained twenty-five thousand plants that could be expected to result in approximately 1,600 *cujes* of tobacco at harvest time. Tobacco farmer and *isleño* José worked alone on his San Pedro-area tobacco farm, which was located just southwest of Havana and produced five hundred sticks of tobacco. A larger labor force, however, did not guarantee a proportional increase in tobacco production. Four additional people labored on the *vega* near Canoa owned by Sebastian el *isleño*, but they produced only one thousand *cujes*. Nineteenth-century manuals for Cuban tobacco farmers indicate that a single *veguero* was capable of tending between 10,000 and 12,500 tobacco plants, but it is uncertain whether monopoly restrictions, land quality, the amount of land available to farm, personal preferences, or a combination of these factors governed the decisions of these particular *isleño vegueros*.[27]

Notary records, particularly wills, provide detailed information about the few individuals described as tobacco farmers and Canary Islanders. On 9 May 1776, just two days before he died, José Antonio González dictated his will in Santiago de las Vegas before notary Carlos Ximénez. The legitimate son of Juan González and Leonarda Falcón, González was born in the Canary Islands. Among the possessions he listed was a *vega* located on property belonging to don José Nicolás Escalera, and approximately 1,200 sticks of tobacco hanging in a barn there. Two African-born slaves–José María Carabali and José Antonio Mina, who perhaps were used as labor on the *vega*—were also listed among González's assets.[28] Having acquired access to both land and slaves in the time that he resided in Cuba, González does not match the image of a poverty-stricken tobacco farmer presented by Ortiz and others.

Canary Islanders formed an important segment of the population in the earliest days of the colony, as noted above, and the largest wave of Canary Island immigration did not begin until 1817, the final year of the Cuban tobacco monopoly. Yet, during the waning days of the Cuban tobacco monopoly, *isleños* do not appear to have been a sizable part of those engaged in tobacco cultivation. Later, after the end of the tobacco monopoly in 1817, Spain embarked on an effort to increase the island's white population. Openly advertising for white immigrants, the crown's *cédula* welcomed not just Spaniards and Canary Islanders, but any Roman Catholic European "subjects of states with whom we are at peace and amity." Foreign immigrants would be granted the same rights and privileges received by natural-born subjects after they resided in Cuba full-time for a period of five years. Spain's goal of counteracting the racial imbalance between whites and blacks in Cuba was a slow one, reached in 1859 when whites again made up the majority of Cuba's population. Over a period of fifteen years—from 1835 to 1850—15,700 Canary Islanders immigrated to Cuba. Approximately 42 percent of all the Spanish-born residents, or 48,552 people, were natives of the Canary Islands by the time the 1862 census was conducted.[29]

The centuries of immigration from the Canary Islands to Cuba prior to the eighteenth century only add to the challenge of identifying the origin of the island's tobacco growers. Those removed by one generation or more from the Canary Islands were most likely integrated into society as Cubans rather than as *isleños*, with no distinction made to describe them as such. Additionally, none of the *vegueros* described initially as *isleños* were listed as such in subsequent *tazmias*. This may indicate that when these individuals were first listed, they were very recent immigrants, and that they had become sufficiently integrated into the community by the time of the next *tazmia* to be known simply by their surnames rather than a descriptive modifier. Ultimately, when Ortiz and others insist that Canary Islanders were a predominant component of Cuba's tobacco producers, they are guilty of an oversimplification.

In all of Ortiz's descriptions of colonial Cuban tobacco farmers, his portrayal of their poverty may have been the most accurate part (although the state of impoverishment was not universal, as he implied). Financial struggles often plagued Cuba's late eighteenth-century and early nineteenth-century *vegueros*, archival documents indicate. For example, in 1772, the captain general threatened to fine *vegueros* east of Havana fifty *ducados* each or about one *peso* for failing to clear the roads with ax and machete within the ordered two months. A local official stated that this would have little effect on the noncompliant farmers, as they did not have the money to pay.[30] The monopoly's own practices often contributed to the economic hardships endured by many of Cuban tobacco farmers. Many routinely complained that they had not received fair prices for their tobacco, or that they had not been paid upon delivery of their crop to the monopoly. When *vegueros* in the Partido de Govea requested that the prices paid for their tobacco be raised, the monopoly refused, but as Minister of the Indies Julián de Arriaga conceded in a letter to Captain General Bucarely, "there was no just motive for it."[31]

By the second decade of the nineteenth century, little had changed, leading one Cuban to contend that the tobacco farmers were being paid only a third of what their crop was worth. Not only were prices unfair, according to the farmers' estimation, the monopoly sometimes failed to pay for tobacco already delivered. Despite these criticisms, little was done to alleviate the situation, and it was suggested in late 1813 that the government require the *vegueros* to turn in their tobacco to the monopoly even if the farmers were not paid at all.[32]

Royal authorities on the island, including Cuba's highest-ranking official, acknowledged the difficulties inflicted on the farmers by these practices. In 1774, the Marqués de la Torre authorized Bayamo's *factor de tabacos*, Martín de Asanza, to pay the *vegueros* for their crop. De Asanza should find money to pay the farmers, the captain general stated, from whatever source possible in order to "satisfy in some way the *vegueros* of this Jurisdiction, who only with the product of their crops can alleviate the misery in which they suffer."[33] By the mid-1810s, at least one official within the Cuban tobacco monopoly itself conceded that promptly paying the tobacco growers for their crop provided a positive stimulus to the industry.[34]

Vegueros in western Cuba frequently cited their inability to provide for their families, yet a closer examination reveals multiple possible causes. While finances may have been a contributing factor, a more pressing one perhaps was related to local supply issues. The rural population west of Havana either raised their own livestock or attempted to buy it from local cattle ranch owners, yet the area's *vegueros* frequently complained about the lack of meat. As early as the spring of 1768, local authorities alerted Havana that the region's *vegueros* did not have sufficient meat supplies to feed their families. The situation remained unchanged by the mid-1770s, despite the creation of a separate jurisdiction and a direct line to the seat of power in Havana through a newly created local government. *Teniente gobernador* (lieutenant governor) Antonio Fernández wrote to Cuba's captain general, the Marqués de la Torre, not long after he arrived in his new post. "I see their unhappiness when they do not find someone in any *hacienda* [ranch] to sell an *arroba* [approximately twenty-five pounds] of meat to them," Fernández informed the captain general. Ranches refused to sell to the local *vegueros*, even though they provided meat to others, as evidenced by the sale of two *arrobas* of meat from *hacienda* San Julian's foreman to a ship that had insufficient supplies to set out on its voyage. Fernández visited Guane del Sur, located on the far southwestern coast, during the Easter holiday of 1775. He reported that the residents there were *buena gente* (good people), but suggested that something should be done to help the poor *vegueros*.[35] Nevertheless, the problem of a lack of meat remained an issue for western farmers even into the next decade.

In 1780, don Nicolás Ramos was the *diputado* (elected representative) for the *vegueros* in Filipinas. That year, Ramos and ten other *vegueros* from the district personally wrote to Fernández and requested official assistance. The farmers indicated their loyalty to Spain via their service to the Crown: "We serve the King . . . with our labors and ourselves. . . . We are ready if the enemy lands on these coasts."[36] The area's newest *teniente gobernador*, Ramón Lloret, forwarded the letter to Captain General Diego José Navarro. Lloret added that some *vegueros* had told him that without meat they were inclined to "leave the country, because it is nothing unusual that they, their wives and children could die from hunger." Lloret lamented, "What am I to do with a *veguero* with a family, who after having gone to all of the cattle ranches and estates of this jurisdiction, carrying his money in his hand (and this, Sir, because I, from my own pocket, knowing his necessity made the charitable act of loaning it to him) with the hope of that someone will sell him a cow or pig, finds no one who will sell him a pound of meat?"[37] Navarro may have been sympathetic to the tobacco farmers' plight, but he noted that he could not make the ranch owners sell meat to the area's *vegueros*, particularly at the established prices. The situation remained unresolved through that fall and winter, and in May 1781, the latest lieutenant governor of Filipinas, Gabriel de Mier, again wrote to the captain general, reporting that local *vegueros* were in desperate need of meat. After pondering the situation, Mier commented that there was only one remedy; namely, the establishment of slaughterhouses offering small livestock in each district seat. This time the request met with a positive response. Navarro gave his approval for the sale of "up to a pound of meat daily at equitable or wholesale prices" for a period of four months. The scarcity of available meat still plagued the area nevertheless as late as 1805. From the San Cristobal River to Mantua, an area covering virtually the entire Filipinas region, there was no butcher from whom the *vegueros* or other residents could purchase meat.[38]

Although Ortiz described some of the social actors responsible for producing Cuba's famous agricultural product, there were many others who cultivated and sponsored tobacco production—including people of color, both slave and free. Archival documents clearly reveal that royal officials recognized slaves as an important source of labor for the island's tobacco industry. During the twenty years between 1740 (its founding) and 1760, the Real Compañía Mercantil de la Habana (Havana Company) sold nearly five thousand slaves "at an average price of 144 *pesos*," with the vast majority (almost four thousand) sold on credit or paid for with tobacco.[39] In 1750, the Crown authorized the sale of up to five hundred slaves at the price of three hundred *pesos* each in cash or tobacco in the hope of increasing tobacco production. While some applauded this move, it was insufficient to meet the demand. Cuba's governor, Francisco Cagigal de la Vega, countered that Havana's districts alone needed 1,500 or more slaves. Enslaved workers were needed on the island not only to sustain but also to increase tobacco production. A drop in tobacco production several years later led the governor of Santiago de Cuba, Alonso Arcos y Moreno, to suggest that the introduction of black labor would result in a larger tobacco crop.[40] Cagigal de la Vega agreed, stating that without slave labor it would be impossible to enjoy fully "the abundance of this fruit."[41] In 1764, the local head of the tobacco monopoly in Santiago de Cuba requested "algunos negros bozales" [a few African slaves] to assist in tobacco cultivation.[42]

Even Spain's King Carlos III (reigned 1759–1788) realized the value of slave labor in the island's tobacco industry, approving the purchase of twenty slaves to assist in the work of the royal monopoly. Despite these measures, not enough slaves were available to fulfill the ever-growing demand. Additional slaves were needed to work in agricultural pursuits, "especially tobacco," Captain General Marqués de la Torre complained to Spain in 1773. One year later, local officials in Cuba continued to petition Havana for more slaves. Bayamo, for example, wanted six hundred or seven hundred slaves to help with planting. The captain general agreed that prompt payment of the *vegueros* and distribution of slaves would contribute to a significant increase in tobacco production, but he informed the manager of Bayamo's monopoly that the funds were simply not available for either.[43]

Spain, however, did see the connection between slave labor and increased tobacco production on the island. The Real Compañía de Comercio de La Habana, the "mixed state and private enterprise" that held the exclusive right over all trade and commerce on the island, promoted the use of slaves in tobacco through discounted prices, special incentives, favorable credit, and payment in kind that benefited Cuba's tobacco farmers. Not only did the Compañía provide the tobacco farmers with a labor force, but it also engaged in a triangular trade involving tobacco and slaves. Tobacco was taken from Cuba to Spain, where it was exchanged and the funds used to purchase cargoes of merchandise needed for the slave trade in Africa. The ships returned to Cuba from Africa, the holds this time filled with human chattel. An unsuccessful proposal made in 1760 sought to strengthen the ties between the enterprise and tobacco. The company headed by José Villanueva Pico offered to "carry free of charge all the tobacco of the Real Factoría from

Havana to Spain, Vera Cruz, or Tierra Firme . . . [as well as] all the cannon and munitions of war from Spain intended for the fortification of Havana, and to bear one-half of the expenses of providing and maintaining a coast-guard to prevent illicit introductions." In exchange, the company sought the right to introduce at least one thousand slaves per year for a period of ten years. This proposal never came about, and a new monopoly was authorized in 1765. It tried one voyage to Africa, which was a complete failure, and then contracted with British merchants to supply slaves.[44]

The tobacco monopoly so valued slave labor that it gave tobacco farmers the funds needed to purchase slaves. As early as 1764, the Real Compañía, the only authorized importer of slaves into the island, brought a shipment of slaves into Havana. All the slaves were distributed to the tobacco growers in the western hamlets at bargain prices. Five years later, the tobacco monopoly distributed slaves to *vegueros* living in the jurisdiction of Havana under the agreement that the farmers would repay the monopoly within two or three years. In 1773, the monopoly again wanted to provide slaves for the *vegueros*, but complained to Spain that it lacked the funds necessary to purchase them. The monopoly's financial situation improved to such a degree by the following year that there now was an "abundance of money" available for paying the tobacco farmers for their crop as well as supplying them with slaves. Between 1 March 1798 and 30 September 1799, the tobacco monopoly provided 2,550 *pesos* to eight different tobacco farmers in Nueva Filipinas in order to assist them with the purchase of eight slaves destined to work in the tobacco fields. This amount represented nearly one-third of the total, 7,860 *pesos*, distributed to all *vegueros* during this period. The *estanco*'s commitment to Pinar del Río's tobacco cultivation was further demonstrated when this amount is compared to the tobacco grown in the region, which was just 7,053 *arrobas* (about 178,794 pounds) or less than 3 percent of the island's entire 1798 and 1799 crop of 252,148 *arrobas* (nearly 6.4 million pounds).[45] Despite the smaller harvest produced in Nueva Filipinas, the monopoly was willing not only to continue to pay higher prices for their tobacco, but also to provide a substantial portion of its available funds to assist area farmers in expanding their labor force. Western Cuba's contribution to the island's tobacco harvest was small, but it was an essential component of its tobacco industry, one in which the monopoly was happy to invest in order to ensure its continuance and success.

These were not isolated cases. On occasion, the *vegueros* personally petitioned the monopoly to request slaves, most notably during the first decade of the nineteenth century. Widow and tobacco grower doña María de la Merced Junco, a resident of the city of Matanzas, sought funds in 1801 for this reason. She wished to purchase two slaves to help with the cultivation and planting of tobacco, she indicated. The following year, the *estanco* provided Güines *veguero* Juan José Pérez with three hundred *pesos* for the purchase of a slave. Two separate cases from 1800 demonstrate that the monopoly considered the *veguero*'s conduct when making decisions as to who should receive the solicited funds. Citing the "good circumstances with which he has distinguished himself," the monopoly granted Manuel Díaz from Güines the money to purchase two slaves. Another *veguero*, don José María Rodríguez from the Partido de las Virtudes, was "one of the good growers, and . . . faithfully carries out his annual deliveries," the monopoly stated when they awarded him four hundred *pesos* to purchase one slave. Nearly fifty years earlier, the cost for slaves in Havana had been 220 to 300 *pesos* each, indicating that most likely the monopoly subsidized the prices that *vegueros* paid for slaves or allowed them substantial discounts, a practice that the tobacco monopoly had utilized before.[46]

The monopoly was willing to assist farmers in providing them with the labor force they requested, but eventual repayment of the debt was expected. Havana-area *veguero* don Agustín Ruiz in the late 1760s purchased two slaves on credit from the monopoly, intending to use the extra labor to increase the amount of tobacco he planted. The terms of the agreement stated that Ruiz would repay the money from his crop within two or three years. As of May 1772, Ruiz had failed to reimburse the monopoly for the 480 *pesos* he owed, and officials believed that Ruiz was hiding in the countryside and did not have a permanent residence. Local official Martín Xavier de Echavarría learned that Ruiz was in the town of Guanabacoa and was planning to escape into the mountains. Echavarría wrote to the captain general and requested that the adjutant major of Guanabacoa's volunteer battalion take Ruiz into custody and seize his belongings in order for the government to recoup its losses. As late as 1792, the monopoly claimed that it had yet to receive payment, although the guarantor of the loan, don Andres Díaz from the Partido de San Geronimo, and Ruiz's son-in-law, don José Calixto Morales, stated they had repaid the money.[47]

Tazmias listed workers on the *vegas* by number, and not by name, making it difficult to determine when slave labor was used, unless the slaves were listed separately as producing tobacco for themselves. Clearly, however, Cuban *vegueros* throughout all parts of the island used slaves in the production of tobacco although at no time did slaves constitute tobacco's primary labor force. Much of the existing historiography of Cuban slavery focuses on slave labor on the island's *ingenios* (sugar estates), yet census data indicates the importance of slaves in other agricultural endeavors as well. A census taken in Güines in 1774 notes that the area was home to 319 male slaves considered to be of prime working age, notably between fifteen and fifty years old. Of those, nearly 77 percent or 245 individuals were found on the district's *vegas*. In contrast, just fifty-eight male slaves between the ages of fifteen and fifty resided on the area's three *ingenios* of Nuestra Señora de la Merced, San Joseph, and Nuestra Señora de Regla. Cuba's sugar estates individually held greater numbers of slaves than did the *vegas*, particularly during the mid-nineteenth century. Yet, the majority of the island's slaves did not work in sugar, even at the height of that industry during the 1860s.[48]

The daily schedule of slaves working in the tobacco fields demonstrated the importance of this valuable labor to the island's tobacco industry. For these slaves and others, their day began promptly at four in the morning with the tolling of the work bell. Slaves in the western tobacco-growing region known as the Vuelta Abajo rose to fetch firewood, water, and feed for the farm's animals, all before heading into the tobacco fields by daybreak. Their morning field duties, which varied depending on the time of year, could include preparing the soil for the seedbed, planting, removing additional shoots from the base of the plant, picking off worms and other pests from the delicate leaves, harvesting the plants, and carrying them to the tobacco barns. From 10 a.m. to 11 a.m. the slave would stop for lunch and a brief rest; he or she resumed work at 11 and continued until dark. At the end of the day, the slaves would return to their huts and eat their last meal of the day, turning in for the night between 10 p.m. and midnight.[49]

In spite of the relative anonymity experienced by much of Cuba's enslaved population working in tobacco, there is documentation of slaves in various parts of the island producing tobacco on their own. This was particularly the case in Cuba's eastern and central districts. The contemporary expectation was that a slave who worked on the *vega* should be capable of handling all aspects of tobacco production, including planting, removing potentially damaging pests such as worms, harvesting, hanging the tobacco in the barn, and preparing tobacco into manageable bundles for transport to the factory.[50] With this vast knowledge base, Cuba's slaves had all the tools at their disposal for producing quality tobacco on their own account and selling it to the monopoly. A 1768 listing of *vegueros* failing to turn in their tobacco included three groups of slaves from the Cauto district near Santiago de Cuba. These slaves, owned by local tobacco farmers Luis Miguel del Río, Felipe Napoles, and don Melchor Delgado, had produced a combined total of one thousand *manojos* of tobacco, but had not delivered them to the monopoly as directed. During the mid-1770s, slaves around Trinidad, farther to the west, grew tobacco and were responsible for independently producing a considerable portion of each year's crop. Seventy *conucos* (plots used by slaves to grow produce or other agricultural products for their own consumption or sale) in the area held 178,000 plants in 1773, approximately 7 percent of that year's crop. The following year, the number of *conucos* significantly increased to 170, accounting for an astounding 1,038,000 plants, or nearly 18 percent of the area's overall total tobacco production. Trinidad's slave *conucos* contained fewer plants over the next two years, 1775 and 1776. Although the number of plants dropped to 139,500 and 39,900 respectively, they still constituted a significant proportion of the area's tobacco harvest. The slaves who grew tobacco on their own were responsible for almost 1.4 million tobacco plants between 1773 and 1776, or in other words, approximately 11 percent of all the tobacco plants in the district.[51]

Slaves regularly grew tobacco on their own account in other parts of the island, as they did in the western districts. One such location was the *partido* of El Cano, located along the Jaimanitas River within the modern-day limits of Havana City Province. In 1771, six slaves grew tobacco for themselves. The slaves belonged to five of the district's *vegueros*, don Julio Capote, Julio Chico, don Manuel de Aragón, don Felipe Torres, and a partially identified farmer with the surname Bello. Most of these enslaved individuals worked alone, each producing between 100 and 250 *cujes*; together, the slaves were responsible for 750 *cujes* of tobacco, or about 5 percent of the district's total recorded crop for that year. The following year, at least seven slaves belonging to five area tobacco farmers produced more tobacco than the previous year, 1,020 *cujes*. Only one of these slaves, the individual owned by de Aragón, may have been the same person who grew tobacco in 1771, indicating that tobacco production may not have been an endeavor that the area's slaves engaged in year after year.[52]

Slaves in Filipinas likewise grew tobacco on *conucos*.[53] Near San Juan y Martínez in 1770, six slaves named Miguel and José Antonio de Otaño, José Antonio de Oropeza, José de Mesa, and Ignacio and Francisco Cordova were listed separately from their *veguero* owners. These slaves worked on four separate *conucos*—probably during the hours between dusk and late night, when they were free after a long day in their owners' tobacco fields—to grow a combined 430 *cujes* of tobacco, or just over 3.5 percent of the district's entire harvest that year. Working in pairs, the Otaños and Cordovas harvested 200 and 150 *cujes* respectively, while Oropeza and Mesa worked alone, each turning in just 40 *cujes* of tobacco (the smallest amounts by far of any *veguero* in the district). Other slaves in Filipinas similarly grew tobacco on their own account. A slave named Pedro, owned by *veguero* Juan Antonio Ledesma and living along the Cuyaguateje River in 1771, produced one hundred *cujes* of tobacco. Although he was responsible for just slightly more than 1 percent of the tobacco cultivated in this area that year, Pedro grew as much or more tobacco than seven other *vegueros*, all of whom were free. Slaves in western Cuba continued to grow their own tobacco into the 1780s. Eight slaves owned by three tobacco growers near Filipinas in 1780 together produced a total of 792 *cujes* of tobacco. When compared to the harvest from the area's most prolific *vegas*, which produced in excess of one thousand *cujes*, this was only a small portion of tobacco produced in the *partido*; however, the combined product of these slave *conucos* was greater than that harvested separately by twenty-six of the forty-nine individual free *vegueros* listed on the 1780 *tazmia*.[54]

Like their enslaved brethren, free people of color also cultivated tobacco. *Tazmias* for the districts of Narajal, Estancia, and San Agustín near Matanzas in 1770 list 25 out of 131 *vegueros*, or 19 percent, as *morenos libres* (free blacks). Almost half—ten—of the freed men growing tobacco had the surnames Contreras or Jústiz. This probably indicates that they had been owned at some point in their captivity (and perhaps emancipated) by either the Marqués de Jústiz y de Santa Ana (Manuel José Aparicio del Manzano y Jústiz) or the Conde de Gibacoa (Jerónimo Espinosa de Contreras y Jústiz). These nobles were local relatives who had earned their titles through their service to the island's tobacco industry. The surnames of two other free black *vegueros*, Miguel Congo and Geronimo Mina, designate them as African-born. *Moreno libre* Juan Manuel Contreras demonstrates the success that some former slaves were able to have in the area's tobacco industry. In 1770, Contreras produced the second-highest amount of tobacco grown that year, two thousand *cujes*, and was the only *moreno libre* out of four people who produced that amount of tobacco in the area.[55]

Free blacks were a significant proportion of those growing tobacco in some areas of Cuba during the late eighteenth and early nineteenth centuries. This was especially the case around Trinidad, located on the central southern coast of the island. An examination of the *tazmia* collected by local tobacco official Fernando Muñoz reveals that just over 20 percent (or 60 out of 293) of all *vegueros* were free people of color in 1772. Four years later, there were more than 120 fewer *vegas* in the Trinidad vicinity. Yet, while the number of tobacco farmers overall had decreased, with the number of white tobacco farmers declining from 233 to 110, the number of free black *vegueros* had grown. There were now 68 free *vegueros* of color, or slightly more than 38 percent of the total *veguero* population of 178. Free people of color did not make up the majority of the area's overall tobacco farmers, but in several locations, particularly along a few rivers near Trinidad, they did. Over time, free people of color were increasingly the majority or near majority of the area's *vegueros*, such as in the district of the Agabama River, where almost 39 percent of the ninety-six *vegueros* in 1773 were *morenos libres* or *pardos libres* (free mulattos). By 1776, the percentage of free tobacco farmers of color had increased to 62 percent, or forty-nine out of seventy-nine.[56]

Yet, people of color, both free and enslaved, were not the only ones who participated in the production of this valuable agricultural commodity. Those who worked to protect Cuba from enemies both abroad and at home likewise made their living cultivating tobacco. Threats from foreign enemies meant that military troops had long served on the island, yet in the wake of the Seven Years' War, the Conde de Ricla (Ambrosio Funes de Villalpando) and Alejandro O'Reilly further militarized Cuba. This increased military presence affected the demographic profile of the *vegueros* in two specific ways. First, retiring soldiers from the regular army received their discharge and some became tobacco farmers. Second, many Cuban residents were now obliged to serve in the militia, as participation became mandatory for most civilian males. In Havana, all men between the ages of sixteen and forty were required to join the militia. For areas with lower populations and therefore fewer potential militia members, the age bracket was expanded to all those between fifteen and forty-five. The height requirement was lowered across the island to 4 feet 11 ½ inches, half an inch lower than required on the Spanish peninsula, in order to extend further the base from which to draw possible militia members.[57] Both Spain and the colonial government in Havana expected the revamped and expanded regular military, together with the citizen soldiers who staffed Cuba's growing militia, to defend the island against any future attack. Whey they retired, these soldiers and militia members contributed to producing Cuba's most well-known export.

Vegueros figured prominently among the new, reorganized militia's ranks. The historic ties between the island's tobacco farmers and the militia date back to at least 1717, with the initial revolt against the tobacco monopoly. *Vegueros* calling for an end to the monopoly were quick to cite their participation in the militia and repeated defense of the island as a testament to their service to the Crown.[58] As the eighteenth century progressed, however, many tobacco farmers protested the obligatory militia service. Royal officials watched them under the auspices of the tobacco monopoly, and as militia members, tobacco farmers endured additional governmental scrutiny while they completed their mandatory service and training. Most *vegueros* astutely argued that militia requirements interfered with their responsibilities on the *vega*, particularly during the tobacco harvest. In 1811, the representative for Holguín's tobacco farmers, Antonio Abad Brito, protested against requiring the tobacco farmers to report for militia service and receive "instructions in the military discipline" every eight days, declaring that the *vegueros* needed this time to work their crops. Local officials cited the *vegueros*' militia service, particularly during times of war, as one of the reasons tobacco production decreased during the late 1790s and early 1800s.[59]

Vegueros did enjoy benefits from militia participation, however, including the ability to take advantage of the *fuero militar* (military privilege) beginning in 1771, a right personally extended to officers and sergeants in Cuba's militia by Spain's King Carlos III. The *fuero militar* allowed Cuban *vegueros* who were also militia members to present their legal cases before a special military tribunal, a right previously granted to only His Majesty's Royal Guard. This honor was more than a superficial nod to those who not only defended Cuba against her enemies, but also produced one of its valuable crops. "In a hierarchical colonial society . . . [it] conveyed special distinction and prestige and reinforced a sense of corporate honor and *esprit de corps*."[60] The *vegueros* and their immediate families were able to take advantage of these benefits, although full enjoyment of the *fuero* was of little use to the *vegueros* when there was no military court to hear their case, as Cuba's captain general acknowledged in 1772.[61]

Mayarí, located near eastern Cuba's north coast, was home to many tobacco farmers who were also members of the local militia. As early as January 1755, Santiago de Cuba's governor, Lorenzo de Madriaga, noted that numerous Mayarí residents, including a lieutenant in the Compañía de Mayarí, don Felix José Rodríguez, worked in tobacco. A monopoly official just over ten years later noted that eighty-one members of the Mayarí district's militia owned tobacco farms, including eleven captains of the white and free colored militias. The *vegueros* were evenly split between the white (forty-five) and the colored (forty-three) companies. The majority, thirty-three, of those assigned to the free black militia were *mulattos*, while ten were part of the *moreno* group. By 1770, only a third of all the militiamen and officers who had owned *vegas* in 1768 remained on the area's *tazmia* of 109 *vegas*. These militia members and *vegueros* owned twenty-six tobacco farms containing 1,016,000 plants tended by ninety-six people, possibly including the militiamen themselves. Most owners, 73 percent, had three or more people working in the tobacco fields, and 63 percent of these had between five and seven *operarios* (workers). Five of the twelve *vegas* employing five or more workers were owned by men associated with the *pardo* militia groups. Four of these men—Luis Miranda, Miguel Felipe Romero, Tomás Barrientos, and Juan Ventua de Pozo—each owned *vegas* with sixty thousand tobacco plants, the first three tended by five people, and the last by six. Miranda and de Pozo knew one another, having served together in the same militia company in 1768 under the command of Captain Vicente Barrientos.[62] The remaining two-thirds of those who had owned *vegas* in 1768 and were members of the militia may have aged out of the requirement to register for duty, left the area, ceased to grow tobacco, or died.

Another tobacco farmer with ties to the island's military was Mateo Morgado, originally from Sancti Spíritus in the eastern-central part of the island. Sometime during the early 1770s, Sergeant Morgado purchased from José Ferrara a tobacco farm located along the banks of the Cuyaguateje River and bordering the *camino real* (royal road). In 1770, a *vega* owned by the forty-four-year-old Morgado and worked by three people, perhaps including the sergeant himself, produced 650 *cujes* of tobacco, nearly 150 sticks more than the area's average, while the following year's harvest was 330 *cujes*, over double the previous year's.[63] By early 1777, he had been promoted to lieutenant junior grade and had purchased a home in western Cuba in the *villa* of Filipinas befitting his new rank. He began to dispose of his *vega* holdings in the early 1780s, selling a tobacco farm located along the Cuyaguateje River—perhaps the same one he had owned for a decade—for 250 *pesos*. A slightly larger property, a *vega* described as capable of being worked by four men, was sold on 27 November 1781 to Rafael Ilobera for the same price.[64] Morgado nevertheless remained actively involved in the area's tobacco industry. In August 1793, the sixty-seven-year-old purchased a farm from Captain (reformed militia) Tomás de Sierra for 150 *pesos*. This *vega* was located along the *Paso Real* (Royal Way) and bordered farms owned by Luis Lantigea and Nicolás Ramos.[65] Mateo

Morgado's actions demonstrate the fluidity within Filipinas's colonial tobacco-growing industry. Its *vegueros*, like those throughout the island, did not buy a farm and hold onto it for decades. Instead, *vegas* frequently changed hands as tobacco farmers searched different areas for the best tobacco land.

Nicolás Ramos was himself a sergeant in the Guane del Sur Company, as well as a tobacco farmer for at least two decades. He also served as the *diputado* (elected representative) of the *vegueros* in the area. Unlike other military members serving in the area, who were predominately from Havana, Ramos was a native of Filipinas. In 1770, the forty-two-year-old Ramos owned one *vega* in the area of Guane del Sur, near the Portales and Cuyaguateje Rivers. The tobacco accounting sheet for that year's harvest noted that Ramos's farm had been worked by four people and yielded one thousand *cujes* of tobacco, making him tied for the second-largest tobacco producer that year. While tobacco production on his *vega* dropped to just six hundred *cujes* the following year, he remained the second-largest grower of tobacco in the area. In 1781, four people continued to work the now-fifty-three-year-old's tobacco farm, but according to that year's *tazmia*, they cultivated only 150 *cujes* of tobacco.[66]

Another *veguero* with military ties was Captain Bernabé Hernández, who was involved in Cuba's tobacco industry over several decades. During the 1770s, Hernández regularly appeared in the *tazmias* for the Xiaraco district. The owner of a *vega* worked by between eight and ten people, Hernández was usually among the largest single-farm tobacco producers in the *partido*. The eight people who worked on his *vega* in 1770 tended 2,200 sticks of tobacco, making Hernández the highest producer in the Xiaraco district for that year. He expanded his labor force to ten people just two years later, and as a result, the amount of tobacco he cultivated increased to 3,600 *cujes*, making his *vega* responsible for the most tobacco grown on any one farm in the area. Only Francisco Hernández and Nicolás Guzman, who owned two *vegas* each, brought more tobacco to the royal monopoly that year. The tobacco industry in Xiaraco expanded quickly during the first few years of the decade, rising from just 86 *vegas*, 283 workers, and 51,290 *cujes* in 1770 to a high of 143 farms, 361 *operarios*, and 118,206 sticks in 1773. Hernández's commitment to the tobacco industry remained strong even as the area's tobacco industry began a period of decline, dropping to 110 *vegas* by 1774. Although his labor force decreased from nine to eight workers between 1773 and 1774, production on his *vega* rose from two thousand sticks in 1773 to three thousand during that period. The number of active tobacco farms in the district dropped to sixty-six by 1775, and continued to decline until 1776, when just forty-nine farms remained, producing 38,750 *cujes*—a decline of nearly 80,000 *cujes* in the span of three short years. In 1775, however, Hernández restored his labor force to ten, a figure he was able to maintain the following year as well, even in the midst of this crisis. By the early 1790s, he had left Xiaraco and moved to Santa María del Rosario, the "private fiefdom" created in the 1730s by the second Conde de Casa Bayona (Francisco José Chacón y Torres), and located to the southeast of Havana. There, Hernández served as the local tobacco farmers' *diputado*.[67]

Captain Nicolás del Pino, born in the Canary Islands, spent many years growing tobacco in the Güines district. A *padron* (census) from 1767 showed ten people in his household, including his wife and three children. *Tazmias* from Güines show that the thirty-eight-year-old del Pino was growing tobacco as of 1768, and by 1771 he owned two *vegas* which were worked by six people, perhaps including slaves. That year, he produced 3,500 *cujes* of tobacco, more than all the other district *vegueros*, with the exception of two individuals. The 1774 *tazmia* listed del Pino with just one farm and a reduced labor force of five people. Two years later, the now forty-six-year-old *isleño* again owned two tobacco *vegas*, indicating perhaps that he had sold one farm between 1771 and 1774, and purchased another sometime between 1774 and 1776.[68]

In addition to members of the militia and the military, individuals from other important groups within colonial Cuban society also participated in the island's tobacco industry, including members of the clergy. Throughout Latin America, the Catholic Church was involved in numerous economic ventures, particularly agricultural ones. As producers of grain, grapes, and sugar in Peru, or as ranchers in Argentina, members of the Catholic Church in the New World expanded its wealth and power through diverse agricultural enterprises. Cuba's Catholic Church was never as wealthy as the church in many of Spain's other colonies. However, as elsewhere, the island's religious orders and individual clergy members were involved in numerous agricultural endeavors, including those of the tobacco industry.[69] One such example was Father Dionisio Manrique, born in Móstales, Spain, to Andrés Manrique and María Delgado y Vargas. Along with serving as the local parish priest in San Felipe y Santiago, near Batabanó, he cared for his young, orphaned nephew (also named Dionisio Manrique) who arrived in the mid-1760s as a fourteen-year-old boy. In 1774, Father Manrique's *vega* produced 3,500 *cujes* of tobacco, making him the third-largest producer of tobacco in the area, behind don Pedro Menocal and Antonio Gutierrez, both with 4,000 *cujes*. Three other farmers each employed larger numbers of workers, but with four people tending Father Manrique's tobacco fields, the priest had more people working for him on his *vega* than 80 percent of the other *vegueros* listed as having *operarios*.[70]

Other men of the cloth also operated *vegas*. In the area around Bayamo in 1773, four separate priests were responsible for farms holding a combined total of 115,500 tobacco plants. Two of the *vegas* were probably about two acres in size. One belonged to don Andres Josef Tamayo, a *presíbtero* (priest) in the district of Bayamo; it had 11,000 plants. Father Santiago Escalante owned the other *vega*, which held 10,500 plants. The tobacco farms owned by Father Felipe Ramírez and Father don Fernando González were significantly larger at about ten acres each, with forty-four thousand and fifty thousand tobacco plants respectively. Interestingly, Father González tied for second among the largest tobacco producers in the district, and Father Ramírez had the third-largest yield. Together, the *vegas* operated by González and Ramírez contained nearly 8 percent of the Casibaroa district's entire crop of 1,267,000 plants.[71]

While the impact of slaves, free blacks, former members of the military, and others has been downplayed in the existing scholarship on Cuba's tobacco farmers, the presence of those in the island's noble class has been completely omitted. Yet, various members of Cuba's titled noble elite participated in the island's tobacco industry. These individuals were involved in virtually every aspect of tobacco, from owning tobacco farms or the mills used to grind tobacco, to donating land to the Crown to be used for new *vegas*. For one man living in Cuba, Spanish-born captain Juan Núñez de Castilla, tobacco was responsible for his ascent to the island's nobility. Núñez de Castilla was born in Almuñecar, Granada, in 1660. He founded the city of San Felipe y Santiago on his estate, Bejucal, with the hope of encouraging tobacco farmers to migrate to the area. The Spanish Crown granted him the title of Marqués de San Felipe y Santiago in 1713 in recognition of his efforts. Later generations of this family also encouraged tobacco cultivation. Nearly one hundred years later, Núñez's great-grandson, Juan Clemente Núñez del Castillo y Molina, the fourth holder of the title of Marqués de San Felipe y Santiago and the first Conde del Castillo (Count of Castile), similarly offered land to be used for tobacco cultivation. Altogether, between 1700 and 1764, six men owed their noble titles to their participation in or ties to Cuba's tobacco industry.[72] While *vega* owners holding titles of Castile were not as common in Cuba as free black tobacco farmers or those with military connections, the participation by the noble elite in growing tobacco further demonstrates that the people sustaining Cuba's colonial tobacco industry were an extremely diverse group that included all levels of Cuba's society.

Other Cuban nobles, like the Marquesa de Cárdenas de Montehermoso (Bárbara Beltrán de Santa Cruz y Aranda), had a more direct association with tobacco and owned *vegas* themselves. Her husband, Nicolás de Cárdenas Vélez de Guevara, was granted the title of Marqúes de Montehermoso in 1764, in response to his defense of Havana during the English occupation. While the marquis referred to tobacco planting as a "pursuit of the lower orders," his wife clearly was not above supplementing her income from such endeavors. The Marquesa de Montehermoso in 1774 owned a tobacco farm in the district of San Antonio, southwest of Havana. Worked by two laborers, the farm produced six hundred *cujes* of tobacco—a moderate amount, considering that the highest amount grown in the *partido* was three thousand *cujes*, and the smallest a mere one hundred *cujes*.[73]

The Marquesa de Montehermoso was not the only Cuban noble with ties to the island's tobacco industry. Francisco José Chacón y Torres, the second Conde de Casa Bayona, also participated in tobacco. Chacón y Torres was born in Havana on 2 April 1712 to cousins Captain José Bayona y Chacón and doña Luisa Chacón y Castellón. The Chacóns were among the first five noble families in Cuba. In 1746, the younger man followed in his father's footsteps, serving as the mayor of Havana. Close family ties further cemented his connections within the island's Cuban-born (creole) elites. Chacón y Torres's uncle was the founder of Santa María del Rosario, and Chacón's niece, Mariana Josefa Tomasa de Herrera y Chacón, who also became Chacón's wife, was the daughter of the fourth Marqués de Villalta, Gonzalo Luis de Herrera y Berrio; the Herreras were another one of the first five noble families in Cuba. Although Chacón's service during the war with Britain was not distinguished, his contribution of thirty thousand *pesos* for repairs and compensation to the dependents of deceased combat veterans further advanced his social standing. Alejandro O'Reilly chose him during the reorganization of Cuba's military to serve as lieutenant colonel of Havana's white volunteer infantry militia, a unit that included many distinguished members of creole society, including a knight of the Order of Santiago.[74]

The Conde de Casa Bayona's vast agricultural holdings included sugar and cattle, and by June 1775 they included tobacco. The *tazmia* submitted by the captain of the Xiaraco district, Bernabé Hernández, noted that the Conde de Casa Bayona owned a single *vega*, located in the *partido* of Xiaraco. The farm's three laborers produced nine hundred *cujes* of tobacco, and while the *vega* was not the largest tobacco producer in the region, it contributed slightly more than 5 percent of the district's total 1775 crop. *Tazmias* from the three previous years do not list Chacón as a *vega* owner, indicating that the farm was most likely purchased in late 1774 or early 1775. The following year, the Conde de Casa Bayona no longer owned a *vega* in the Xiaraco district. Yet, as noted earlier in the discussion on western Cuba, turnover among those engaging in tobacco cultivation was common. In Xiaraco, for example, 216 individuals between 1772 and 1776 owned *vegas*, a number far greater than the maximum number of farms, 143, present in any one year. Even more revealing, only 15 *vegueros* out of 216 consistently produced tobacco every year during this period, further demonstrating the often-temporary ownership of tobacco land.[75]

While the Conde de Casa Bayona's quick entry into and exit from the *partido*'s tobacco-growing industry mirrored the overall patterns found elsewhere, it may have been influenced by other factors more specific to Xiaraco. Tobacco in the Xiaraco district reached its apex in the early 1770s. There were 99 tobacco farms in 1772, and the following year, this increased to 143, the highest number of *vegas* in the *partido* at any time between 1772 and 1776. Yet, by 1774, the area's tobacco industry entered a period of decline, dropping from 110 tobacco farms to 66 in

1775. Only forty-nine *vegas* remained in the entire district in 1776. The number of people owning tobacco lands also changed during this time. While 140 individuals owned *vegas* in 1773, the highest number during this period, by 1776 only 46 people held tobacco farms. The impact on those laboring in the fields was even more dramatic, which experienced a decline of more than a third to just 101 workers overall in 1776. The *partido*'s tobacco production followed a similar downward trend, producing 60,850 *arrobas* of tobacco at its peak in 1772. Crop harvests declined over the next two years, dropping to around fifty-nine thousand *arrobas* in 1773, and about forty-two thousand *arrobas* in 1774. In this context, it is clear that tobacco in the district was on what would become an irreversible downslide at about the same time that the Conde de Casa Bayona most likely purchased his *vega*. Between 1775, when production was a mere 17,642 *arrobas*, and 1803, the *partido* of Xiaraco failed to see tobacco production exceed 28,000 *arrobas*. This severe drop was not a result of lower prices paid by the monopoly, however, as the prices paid for the highest grade of tobacco in Xiaraco increased. Between 1763 and 1785, the monopoly paid thirty-five *reales* per *arroba*, and increased the price to sixty *reales* per *arroba* by 1803. Rather, the district's shrinking tobacco harvests were due to a devastating hurricane cycle.[76]

Other Cubans holding titles of Castile were even more heavily involved in Cuba's tobacco industry. As early as 1775, Spain's minister of finance, Miguel de Múzquiz, recognized two particular Cuban nobles as the "principal tobacco growers" in the Matanzas jurisdiction. Cousins, the Conde de Gibacoa and the Marqués de Jústiz owned a tobacco mill in Matanzas. The pair was contracted by Spain as early as the late 1760s to produce ground tobacco in the district. In 1769, they were granted an extension of five additional years on their original contract, but the following year the terms were reduced to just three years. These nobles, like other wealthy Cubans, were involved in various economic pursuits, including cutting wood, which officials accused them of doing without the proper licenses.[77]

Cuba's noble elites also participated in the tobacco industry by either donating or selling land to the Crown for the sole purpose of growing tobacco. The fourth Marqués de San Felipe y Santiago who also held the title of the Conde del Castillo (Juan Clemente Núñez del Castillo), and the Conde de Mopox (Joaquín Beltrán de Santa Cruz y Cárdenas) all became involved in this way around the turn of the nineteenth century. The Marqués de San Felipe y Santiago donated 100 *caballerias* (about 3,300 acres) of land. At least thirty-four individuals received tobacco land, varying in size from one *caballería* (approximately thirty-three acres) to twenty-one *caballerías*. Not all the land donated by Cuban nobles for growing tobacco was viewed as suitable for that purpose, however. When the Conde de Mopox donated 294 total *caballerias* (approximately 9,700 acres) along the river to the monopoly, two witnesses provided written testimony that only 94 *caballerias* (just over 3,000 acres) were appropriate for growing the plant. Some argued that this should invalidate the transaction, but it did not, and the contract stood. Others countered that the land was indeed good; nevertheless, the monopoly's board decided against purchasing additional land.[78]

The Conde de Mopox's problems with Cuba's royal tobacco monopoly continued. Contracted to provide tobacco to the monopoly, he delivered the 1804 crop, offsetting his debt by 14,618 *pesos*.[79] The following year, however, after the 1805 crop was processed, his balance due increased to 33,412 *pesos*. Government officials were not concerned, even though the count owed a large debt to the Spanish treasury. "The Conde de Mopox y Jaruco has just given me a positive test of his love and loyalty to the King, and of his predilection to be interested in the tobacco monopoly," a royal official noted.[80] The count would not make restitution for this debt, however. In late March 1807, the head of the tobacco monopoly in Havana, Rafael Gómez Robaud, informed Manuel Godoy that their mutual friend was gravely ill.[81] The count died in Havana on 5 April 1807 at the age of thirty-seven years and six months, leaving an outstanding debt to Spain's treasury of over a million *pesos*, but as Allan J. Kuethe notes, "this debt was forgiven his heirs!"[82]

Most of the individuals described above as associated with tobacco cultivation and production were men, yet the presence of women in Cuba as promoters of tobacco cultivation dates back to at least 1760. That year, Ynes de Arvelo from the Canoa district was listed in a survey of individuals who had turned in leaf tobacco. Like their male counterparts, the females involved in the tobacco industry were from various racial and social backgrounds. Social codes strictly controlled the lives of some of the island's women. Those considered to be of respectable social standing found their conduct dictated by societal mores, which forbade proper women from walking more than two blocks on Havana's streets, and required them to remain in near-seclusion. Activities such as visiting other women, reading, peering out from behind the iron-barred windows of their homes, going on carriage rides, or attending church were the only acceptable ways for these women to spend their time. These same rules of conduct, however, did not apply to women of lower socioeconomic standing, including free blacks or slaves.[83] New research has begun to establish the personal agency exercised by Cuban women in financial transactions despite these societal constraints. In addition to the women from western Cuba described above, María Asuncíon de los Reyes and María Loreto, both of whom owned tobacco land and were married to tobacco farmers, women elsewhere in Cuba participated in the island's tobacco industry. Near Guanabacoa, doña Bernabela Piloto's four workers cultivated five hundred *cujes* of tobacco in 1770, while Gertrudis Oliva rented a *vega* in the *partido* of Govea. Her three workers produced one thousand *cujes* of tobacco in 1774. Widows were also found among the diverse ranks of the island's *vegueros*, including Matanzas-area residents María Antonia and María Lopez in 1770, and doña María Perez the following year.[84]

TAZMIAS PROVIDE ONE of the best windows into the composition of those owning or renting land on which tobacco was grown. It is these period documents, along with other sources, that demonstrate the diversity among Cuba's *veguero* population. People of color, both slave and free, worked alongside their white counterparts in the island's tobacco industry during the mid-eighteenth and early nineteenth centuries. Additionally, various members of colonial Cuban society, ranging from members of the military to clergy and noblemen and women, sponsored or directly participated in the growing of one of the island's most important agricultural crops.

The diversity of those growing tobacco was not an anomaly, nor was it limited to Cuba. *Vegueros* in Mexico also covered a vast spectrum of that colony's society, with poor farmers and Native Americans along with the wealthy all growing tobacco. Faced with the prospect of signing a contract with unacceptable prices, one of Mexico's most prominent planters offered to expand his production allotment if he would be paid based on existing prices for monopoly tobacco. He proposed to plant between twelve and thirteen million tobacco plants, an amount of tobacco that would cover approximately 2,400 acres. His allotment at the time was 6.5 million plants, which was already significantly higher than that of most of the other *vegueros* in the area, who each tended between 50,000 and 300,000 plants. Mexico's titled nobility also participated in tobacco. The largest tobacco landholders in Orizaba and Córdoba, located southeast of Mexico City, were three titled nobles, all absentee landowners who rented out their properties.[85]

The description of Cuba's colonial tobacco farmers as poor, white, and Canary Islanders was a simplified representation of the vast array of people involved in tobacco cultivation. Despite the variety in their backgrounds, socioeconomic rank, or even status (with slaves as well as free individuals earning income from tobacco), many people across all levels of colonial Cuban life assisted in bringing the island's most famous agricultural product to the monopoly. The frequent turnover of those owning tobacco lands only helped to increase the number of individuals directly affected by Spain's tobacco monopoly.

4
"A matter of chief importance"

EUROPEANS WERE INITIALLY uncertain about the value of the weed consumed by New World inhabitants. Their primary economic focus centered on searching for those commodities they knew had a high value in Europe—gold, silver, and other precious metals. This was not an easy task in the Caribbean, but they found much success when Cortés and Pizarro overcame the Aztec and Incan empires respectively. Over time, tobacco gained in popularity and acceptance among Europeans in the New and Old Worlds alike. Tobacco consumption began to skyrocket during the century following the Columbian Encounter, and European royal houses realized the potential for cashing in on this new habit. Spain had once overlooked this New World plant, but tobacco now ranked high as part of the peninsula's financial strategy.

Tobacco expanded its influence over Cuba, Spain, and the Atlantic world, this time through economic means, via monopolies. Various forms of tobacco monopoly sprang up throughout the countries on the continent and their overseas colonies during this period. Even once-outspoken critic of tobacco England's James I and the Catholic Church joined in this new acceptance of tobacco.[1] James I acknowledged the profit potential from the sale of tobacco and in 1613 took over the sole importation of tobacco. Despite previous edicts to the contrary, Pope Benedict XIII in 1725 allowed the use of snuff inside St. Peter's Basilica. Fifty-four years later, the papacy sought to capitalize on the economic possibilities of tobacco and opened its own factory.[2]

The island's tobacco monopoly, part of the Bourbon Reforms, also served as a method of economic control over Cuba's residents, particularly those involved with the cultivation of tobacco. Designed to increase revenue to the Spanish Crown, it also placed tobacco growers and sponsors deeper under the control of the monarchy. Spain and its bureaucratic entities determined where and how much tobacco planters could grow. Additionally, the monopoly gave only the Crown the right to purchase tobacco, prohibited farmers from selling the surplus, and set the official prices for the purchase of tobacco. Combined, these measures strengthened the economic ties between Cuba and Spain.[3]

Spain, as shown elsewhere in this work, could turn to various locations for tobacco if transatlantic shipping faced disruptions or if natural disaster devastated Cuba's tobacco crop. Other sources might not be preferable, but they were better than nothing. The tobacco habit reached deep into Spanish culture and society, and when pressed, the nicotine addict did not care where his or her product originated. For Cuba, however, particularly during the late eighteenth and early nineteenth centuries, tobacco was its primary export crop. Tobacco growers and sponsors had little recourse against the monopoly. The initial response to the monopoly's implementation was outright rebellion. Other options available to the farmers included turning to contraband, destruction of the tobacco crop, or other subtle methods of resistance.[4] There were numerous risks to entering the contraband tobacco trade, however, and some may not have been willing to undertake them.

Spain's earliest approaches to tobacco control at home and in its overseas colonies focused on attempting to limit the plant's availability as a contraband product. The best way to combat this popular illegal trade, Spain decided, was to prohibit all tobacco cultivation for ten years in Santo Domingo, Cuba, Margarita, Puerto Rico, Cumaná, and Nueva Andalucía, starting in 1606. Royal officials hoped that eliminating production would prevent smuggling. England, Spain's European rival, around this same time began to expand in the New World and entered the tobacco trade with its own colonial product from North America. England too sought to protect its imperial interests, creating taxes on foreign tobacco in 1614 in the hope of steering consumers to the product from its own colonies. Spain's Felipe III (reigned 1598–1621) reversed the 1606 prohibition on tobacco cultivation that same year. Scholar Laura Náter notes, "With this move he hoped to force the English to go to Seville and pay high prices for the most popular tobacco of the era." The Spanish monarch was willing to open his empire to tobacco growers. Smugglers, however, did not fare so well under the new regulations. For those who chose to evade the law, the death penalty awaited.[5]

The tight government control encouraged only more smuggling, unfortunately. The governor of Cumaná, located in modern-day Venezuela, lamented the situation just one year later in a letter to the Crown. "Tobacco has been one of the chief factors causing these coasts to be so much frequented by pirates, by reason of the great demand there has always been for it; and in truck for it they give merchandise that comes cheaper than if it were bought in Spain," he wrote.[6] Officials determined that more stringent measures were needed by the 1630s. These regulations, however, were not merely a response to illicit trade, but instead attempted to capture the profit available from selling tobacco through a state monopoly. In 1636, Spain established an official tobacco monopoly in Castile and Aragon, making it among the first European nations to control the tobacco industry. Officials devised the monopoly to add much-needed funds to royal coffers. Nevertheless, it was an unexpected choice, at least from a juridical point of view. Tobacco historian Marcy Norton argues, "Contemporary jurists viewed royal monopolies as an illicit way for sovereigns to raise revenues, since they were consumption taxes in disguise (which required the consent of the Cortés [Spanish legislature]) and unfairly raised prices on basic necessities."[7]

The greatest change to Spain's agricultural and economic policies toward tobacco came with the turn of the eighteenth century. The death of Spain's last surviving Hapsburg monarch, the heirless Carlos II (reigned 1665–1700), led to the ascension of a new royal line to the throne. The Bourbon monarchs were stronger and more effective at governing than the Hapsburgs. The Bourbons implemented numerous economic, military, administrative, and cultural reforms during this time. One such measure included extending the royal monopoly on tobacco to Spain's overseas empire. As would be the case with many other Bourbon reforms, Cuba served as the testing ground for Spain's colonial tobacco monopoly. The Spanish Crown eventually applied the Cuban model elsewhere in the Spanish Empire—to Peru (1745), Chile and La Plata (1753), Santo Domingo (1754), Mexico (1765), Guatemala (1766), Colombia (1778), Venezuela (1779), and Puerto Rico (1785). The document creating Mexico's royal tobacco monopoly in 1765 stressed the economic benefits to the Crown. The Mexican *estanco*, the decree stated, was designed to be a revenue-making venture (the Philippine Islands' monopoly created in 1769 was likewise charged with this goal). For Mexico, whose financial obligations included supplying Cuba with a yearly *situado* (transfer of royal funds), the monies were needed to cover the increasing costs of defending the Spanish Empire, an objective that had become even more important since the 1762 loss of Havana to the British.[8]

As time passed, more and more tobacco entering Spain came from Cuba. One Spanish monopoly official noted that the amount was approximately five hundred thousand to six hundred thousand pounds of Havana leaf per year. Cuban tobacco in the 1690s accounted for 83 percent of all tobacco imports into the Spanish port city of Cádiz, up significantly from only 2 percent at mid-century. By the end of the seventeenth century, tobacco brought more money into Spanish royal coffers than any other source, including New World bullion. This close correlation between tobacco and Cuba led to the island's place as the most significant participant in Spain's tobacco trade, so much that by the early eighteenth century, tobacco was one of the island's primary export items.[9]

Cuba, however, produced no more than three million pounds of tobacco in any one year prior to the creation of the island's tobacco monopoly. A review of Cuba's tobacco industry nonetheless led royal officials to believe that up to 7.3 million pounds of tobacco—worth between six and seven million *pesos* in revenue to the Spanish treasury—were possible. Most of this anticipated increased production in tobacco would be destined for Spain and its colonies, with an estimated 1.5 million pounds available for foreign export. On 11 April 1717, Spanish king Felipe V (reigned 1700–1746) created a royal monopoly on Cuban tobacco. Spain needed a guaranteed influx of Cuban tobacco. Without it, Spain would have to purchase it elsewhere, the king argued.[10] The monopoly, organized under the name Real Factoría de Tabacos (Royal Tobacco Factory or Monopoly), would be administered on the local level by agencies. These entities, known as *factorías*, were to be established at four locations in Cuba (Bayamo, Trinidad, Sancti Spíritus, and Santiago de Cuba), and six locations elsewhere in the empire. Monopoly officials would categorize, grade, and price tobacco according to its quality. In an attempt to circumvent any potential opposition to the new monopoly, official prices for tobacco would be set to those similar to what tobacco farmers had previously received.[11]

Spain's new policy meant that the Crown would exert total authority over all aspects of Cuba's tobacco industry. While this in many ways exemplified the type of economic and social control embodied within future Bourbon reforms, this was unprecedented for Cuba, and indeed for Latin America. The monopoly marked a significant change in the way that Cuban tobacco farmers and sponsors conducted business. They had been stripped of all economic independence. No longer did they have the ability to sell what they correctly deemed as an exclusive and highly valuable product to the buyer of their choosing. From where tobacco could be cultivated, or not, to the amount of tobacco grown and the price paid for it, all the decisions had been taken away from them. The tobacco farmers' response was swift and eventually turned violent, a reaction Spanish officials had not anticipated.

Cuba's *vegueros* resented state control of their industry, and challenged the new economic measures by appealing to local officials and calling for an end to the monopoly. At a 6 August 1717 Havana town council session, the *cabildo* denied the farmers' petition. Seventeen *vegueros* protested this decision, arguing that they did not deserve these restrictions on their industry, given that they were loyal subjects of the Crown who had repeatedly defended the island as militia members. They also addressed the economic impact these measures would have on those who grew tobacco. Many tobacco farmers were very poor, the *vegueros* contended, and did not even own the land they worked. Surely the *cabildo* would take these factors into consideration, the farmers said. The monopoly remained in place despite their pleas, however, and the *vegueros* were informed that only the governor could remove it.[12]

The *vegueros*' lines of reasoning exhibited respect for the Crown, but also revealed their belief that the metropolis had selected them unfairly for this new measure. The unspoken implication of their economic position was that tobacco farmers in general, particularly those whose sole livelihood rested on the product they grew on rented land, would be unable to support themselves without access to the open market for their products. Under the traditional practice in place at the beginning of the eighteenth century, commercial merchants paid Cuba's tobacco farmers for their crop in advance. The new tobacco monopoly provisions changed this, instead paying *vegueros* when they presented their tobacco to the monopoly at late-summer purchasing fairs. The farmers' resulting financial burden of having to self-finance the production of their tobacco until the point of its sale to the monopoly (rather than receiving the income in advance, under the old practice) could endanger their families' well-being. Some might even be forced out of tobacco cultivation altogether. These factors made the new economic measures especially worrisome to the farmers.

Public outcry over the tobacco monopoly increased over the next several weeks. By attempting to seek relief from local officials against the tobacco monopoly, Cuba's *vegueros* had demonstrated respect for governmental institutions and colonial representatives. Throughout much of August 1717, public opposition to the tobacco monopoly increased. Before dawn on 21 August, armed groups of Cuban *vegueros* gathered south of Havana at Jesús del Monte and marched northward, encircling the capital and preventing reinforcements from reaching the governor and his soldiers. Facing no opposition, the growers entered the city and demanded that captain general and governor Vicente Raja renounce his position. Raja did just that, on 24 August, returning to Spain along with the four bureaucrats charged with implementing the new monopoly. This episode marked the first violent expulsion of a high government official in Cuban colonial history.[13]

The *vegueros* viewed the captain general's exit as a total success. Raja's departure made enforcement of the new monopoly virtually impossible, and the conspirators were offered a full pardon. A new captain general, Gregorio Guazo Calderón, arrived in 1718, accompanied by one thousand soldiers; only then did the government reestablish the monopoly, with some minor concessions to the farmers. A royal order, dated 16 December 1718, named Manuel de León y Navarro as the monopoly's director general. The *factorías* were reinstated in the same locations as before. Hoping to prevent another uprising, royal officials gave the monopoly instructions to have on hand sufficient funds to pay the tobacco farmers for their crop.[14]

Tensions over the monopoly remained between the government and Cuba's tobacco farmers, however, and in June 1720, violence again erupted. More than two hundred tobacco growers revolted, and for two weeks they burned tobacco houses and crops belonging to nine residents around Santiago de las Vegas who the growers believed were inclined to accept the monopoly terms.[15] At first glance, the violence directed against the very crop the *vegueros* grew seems disproportionate and unexplainable. Taken in historical context, however, it is clear that the island's tobacco growers knew that dealing with colonial officials offered little hope for change. They had attempted that very route previously and made insufficient progress. The only responses left to Cuba's *vegueros* were economic measures of resistance, such as destruction of the crop clearly valued by the Spanish, and punitive measures against those who had accepted the monopoly's terms. The government responded to the fortnight of violence by detaining the individuals believed to be the principal leaders of the movement. The *vegueros* subsequently avoided a direct attack on Havana, instead laying siege to the city by damming its aqueduct and preventing food deliveries. Spain attempted to address the tobacco farmers' concerns and made some small concessions to the farmers. Beginning in late October 1720, the Crown allowed Cuba's *vegueros* to sell freely any surplus tobacco remaining once the monopoly met its needs.[16]

Over the next two years, Cuba enjoyed relative calm. The tobacco farmers' primary complaints with the island's tobacco administration reflected the economic implications of the *estanco*. Yet, not all *vegueros* fought for the same objectives; personal finances drove their particular economic priorities. The poorest farmers asked to sell their crop at slightly higher prices directly to the urban cigar and cigarette factories. Other, more affluent *vegueros* demanded the governor's expulsion from the island.[17] While these requests may not have seemed related, they were. For those for whom tobacco constituted the economic link for survival, the ability to increase revenue yet still comply with monopoly regulations was sufficient. Those with more financial means resented outside economic control, and responded with demands that centered primarily on a political response. Even these wealthier tobacco growers, however, exhibited an economic reaction to the continued policy. The governor represented the highest colonial authority on the island, and with Raja's departure in 1717, some may have believed that subsequent administrations might offer a more lenient approach toward the island's inhabitants and their economic pursuits. In essence, what the more affluent *vegueros* sought was a more laissez-faire approach to their industry. The opportunity for the farmers to sell their crop at higher prices within the monopoly may have been welcome, but a dramatic change in policy was what they desired most.

The tobacco growers' demands went unanswered. In response, they threatened to not plant tobacco until prices increased, and to burn all the snuff stored in Havana, including any found in the royal stores.[18] Whether they were wealthy or not, Cuban tobacco growers realized the addictive and economic power of their product. Their failure to plant tobacco would lower the supply and surely drive up prices. The destruction of snuff in Havana would create severe scarcities of this popular tobacco product, encouraging local consumers to protest. The farmers may have thought that the public would eventually support the *vegueros* in their struggle against the monopoly, which would aid their cause. A shortage of the popular Cuban tobacco would force the Spanish Crown to submit, resulting in a positive economic outcome for the *vegueros*. Cuban tobacco farmers, regardless of their particular socioeconomic background, realized the possibilities raised by the economic power they possessed as the producers of one of the island's best-known and most popular exports.

The third and final *veguero* insurrection took place in early 1723. Unlike previous revolts, this one saw the tobacco farmers split along socioeconomic lines. Some refused to stop planting tobacco, arguing that it was their only way of supporting themselves and their families. A group of approximately three hundred farmers converged on San Miguel del Padrón, which was located just southeast of the southernmost tip of Havana's bay. The angry *vegueros* ripped up tobacco plants, first from their own fields, and later in nearby plots in early February. Many *vegueros* additionally agreed not to plant the crop for two years, choosing to wait instead for higher prices. Governor Guazo Calderón warned the tobacco growers that the state intended to punish dissenters with death. The farmers remained undeterred. Residents of towns located farther away from the capital, San Felipe and Bejucal, sought the governor's help, and he responded by sending soldiers to nearby Santiago de las Vegas. On the morning of 20 February, soldiers prepared to face off against a reported six hundred *vegueros*. The resulting skirmish saw one soldier wounded, and another was thrown from his mount after the animal's reins were cut. Numerous *vegueros* were injured, and the government soldiers captured eleven men, forcing the remaining farmers to retreat.[19]

The next evening, one hundred armed men gathered in Guanabacoa, located just to the east of the capital. The group was intent on rescuing the imprisoned *vegueros*, three of whom were from Jesús del Monte, three from San Miguel del Padrón, three from Guanabacoa, one from Havana, and one described as a vagrant who lived in the woods.[20] A local priest was able to calm the group, which quietly dispersed the following morning. The governor, fearful of more violence, resolved to teach the rebels a lesson. He ordered the eleven captured *vegueros* executed in the early morning hours, and insisted that they not receive religious rites. As a warning against further revolts, the *vegueros*' quartered bodies hung in trees along the roadside for forty hours while soldiers stood guard. Only after pleas made by Bishop Valdés, the first Conde de Casa Bayona, and others did the governor acquiesce to having the corpses taken down and released to their families. Eight of the men were buried in the nearby church in Jesús del Monte.[21]

Cuba's *vegueros* now realized the precarious position in which they found themselves. The earlier revolts had allowed them to demonstrate their displeasure with the new economic controls of their industry. The island's tobacco farmers additionally had experienced some earlier measure of success. On several occasions, their actions had resulted in lessening—or at least postponing—the establishment of the tobacco monopoly's provisions. Yes, the *vegueros* were the growers of a popular export, but substitutes were available for the Spanish to purchase. The power over and final economic control of the island's tobacco industry ultimately rested in the hands of the Spanish Crown and its colonial representatives. Cuban tobacco farmers could object to the new regulations, threaten to sell their tobacco to non-state buyers, and engage in fraud against the state monopoly, but the *estanco* would remain in effect. For the *vegueros* considering such options, stiff penalties and even the possibility of death, as in the case of the eleven imprisoned tobacco farmers, loomed large. The violent reactions by the Cuban *vegueros* to what they felt were unfair governmental controls on their industry were not without historical precedent. Popular uprisings against colonial authority in the New World often were responses to measures taken against tobacco. Some historians cite the displeasure of Virginia's tobacco growers with colonial taxation, particularly the tax on tobacco, as a mitigating factor in Bacon's Rebellion (1676). Just six years later, the government's control of tobacco again prompted Virginians to respond with violence. Challenging measures implemented in the 1680s to ensure a stable and high price for Virginia tobacco, farmers set out on a tobacco-cutting riot in May 1682. The revolt began in Gloucester County and spread to New Kent, Middlesex, and York Counties. Farmers first destroyed their own crops, as was later the case in Cuba, and then moved on to neighboring farms. An estimated ten thousand hogsheads (five million pounds) of tobacco were destroyed in Gloucester and New Kent Counties alone by the end of the month.[22] This would not be the last time, however, that Virginia's tobacco farmers resorted to violence.

In response to the Virginia Inspection Act of 1730, which required that all tobacco be subject to inspection by public officials, tobacco growers again rebelled and refused to bring in their tobacco. Four warehouses, in Virginia's Prince William, King George, Northumberland, and Lancaster Counties, were burned in March 1732. The violence spread to Maryland, where wealthy planters joined the uprisings. Laws were on the books in Virginia that made destruction of a tobacco warehouse an offense punishable by death, but colonial officials were reluctant to make martyrs out of those responsible and feared reprisals if such action was taken.[23]

Virginia's tobacco farmers, moreover, were not the only ones who responded with violence when confronted with controls they perceived as excessive. Tobacco monopoly opposition triggered at least three similar revolts against the Spanish Crown during the latter half of the eighteenth century alone. Chileans rebelled against official controls on tobacco in 1776, while Venezuelan growers in Grita reacted similarly just three years later. New Granada's 1781 *comunero* revolt also had its roots as an opposition to the region's tobacco monopoly. John P. Harrison observes that, as had been the case in Cuba, "the treachery and deceit with which the *comuneros* were put down served to fortify in the locality the conception of the tobacco monopoly as the epitome of Spanish tyranny."[24]

Cuba's first attempts at a tobacco monopoly disappointed Spain. On 17 June 1724, the monopoly's activities in Cuba were suspended. Not only had the farmers reacted violently to control of their industry, the tobacco sent to Spain was of poor quality and not enough to fulfill the monopoly's needs. Cuba's best-quality tobacco meanwhile was smuggled off the island and sold elsewhere. This suspension of the *estanco* was short-lived, however. By 1727, the monopoly had been reorganized and given the name Intendencia General de Tabaco (General Intendancy of Tobacco). Its first head was Martín de Loynaz, a high-ranking Spanish official formerly in charge of the *renta de tabacos* (tobacco monopoly) in Spain. The period from 1725 to 1733 was one of "relatively free trade"; however, the contraband traffic in Cuban tobacco expanded, and the island was unable to provide the requested three million pounds of tobacco per year.[25]

Spain farmed out the monopoly by 1734, hoping that private leadership would improve the situation. Cádiz businessman José Tallapiedra ran the monopoly from 1735 to 1738. Its management then shifted to Francisco Sánchez de Madrid, the Marqués de Casa-Madrid. The creation in 1741 of the Real Compañía de Comercio de la Habana (Havana Company), which oversaw not only tobacco but also other agricultural products, again changed the administration of the tobacco monopoly. An investigation of the company's practices during the 1750s, however, revealed that the Real Compañía had violated the terms of its charter and presented false statements to authorities in 1746 and 1747. As a result, the Real Compañía lost its right to the tobacco trade, and the monopoly was again reorganized in 1761, this time as a government agency known as the Nueva Factoría de Tabacos de la Habana (New Tobacco Factory of Havana).[26]

When Havana fell to British forces during the Seven Years' War (1756–1763), Spain sought to protect its economic interest in the plant. Included in the articles of capitulation was a provision regarding tobacco. Spain requested the right to keep tobacco currently in the monopoly's possession, as well as other tobacco grown in Cuba. The plan was, Spain hoped, to continue with the monopoly as before, but it was thwarted. While Britain agreed to allow Spanish soldiers the right to leave the island with their equipment and money, it would not allow the exportation of tobacco. José Rapún, the monopoly's head, remained in the city, but the *estanco* had little control over tobacco, and the tobacco monopoly *junta* (board) did not meet. The British occupiers seized 244,557 *pesos* belonging to the tobacco monopoly, and purchased tobacco from growers at double the price paid by the Spanish; however, little changed for farmers outside of the British-occupied areas. An especially forward-thinking Cuban tobacco official had sent 174,748 *pesos* to the central and eastern parts of the island to purchase that year's crop before the siege began, ensuring their continued compliance with Spanish regulations.[27]

With Cuba focused on the British, nearby tobacco-growing regions saw an opportunity to promote their own tobacco to Spain. Santo Domingo's governor, Manuel de Azlor y Urríes, sent seventeen loads of various grades of tobacco to Spain in the hope of stepping into Cuba's spot in the tobacco market. The Crown rewarded the governor's effort by creating a tobacco monopoly administration district for the city of Santo Domingo by mid-October 1763. Meanwhile, the Cuban capital returned to Spanish control under a new captain general and the monopoly resumed its functions. Santo Domingo's hopes of occupying Cuba's prized position as one of the eminent producers of the empire's tobacco were shattered.[28]

Cuba's tobacco monopoly characterized the economic control exerted by the Bourbon reforms in several fundamental ways. In theory, if not always in practice, the monopoly was responsible only to the Spanish Crown and its designated representatives. The monopoly's duties were three-fold—first, to ensure that the tobacco mandated by the *estanco* entered the monopoly, and that no illegal contraband or tobacco production took place; second, to pay the farmers for their product; and last, to send the tobacco to Spain as directed. Each of these functions comprised many components, but each contributed to the Spanish Crown's overall control of the tobacco industry. After the 1717 implementation of the monopoly in Cuba, Spain's preoccupation with controlling the tobacco industry soon extended to its other colonial holdings. Spain extended royal tobacco monopolies to virtually all of its colonies capable of growing tobacco beginning in the 1740s and continuing for the next several decades. In 1788, 858 farmers in Costa Rica grew 10,435,000 plants, producing 41,088 *arrobas* (approximately 4.1 million pounds). The distant Philippine Islands sent over 13.6 million pounds of tobacco annually to Spain from its own royal monopoly as late as the 1860s, when it was one of Spain's few remaining colonial territories.[29]

Spain's colonies, particularly Cuba, ultimately accomplished the goal of creating a relatively secure stream of tobacco into Europe, which funneled much-needed funds into the peninsula's coffers, but not all European-sponsored tobacco monopolies flourished. This was especially the case in Louisiana where, despite their best efforts, both France and Spain abandoned tobacco production under a monopoly. From the beginning, France was uncertain about tobacco's viability in Louisiana as a successful agricultural commodity. Its cultivation was possible, but there were other significant players in the tobacco market, specifically Cuba and Virginia, where tobacco had a greater economic potential. The area's tobacco prospects seemed doomed by 5 September 1730, when the new lessees of the French tobacco monopoly, the United General Farms, were required by the Indies Company to buy Louisiana tobacco at the same price paid for Virginia tobacco, even though it was much more expensive to produce. Transportation costs alone made it difficult to compete with Virginia tobacco, and if the monopoly had attempted to do so, the French colony's farmers would have received very little compensation as the profit margin would have hardly made Louisiana tobacco worth its while. Nevertheless, the monopoly continued to hope for profitable harvests, but by the early 1760s Louisiana's farmers were still producing only enough tobacco to fill one ship per year—no more than they had in 1731. With the cession of the territory to Spain in the 1760s, France's hopes for tobacco cultivation in Louisiana ended. Instead, France shifted its plan for an overseas tobacco source in the New World to its Caribbean island holdings in Martinique and Guadeloupe, but they too soon were abandoned, as tobacco cultivation proved impractical in those areas as well.[30]

Spain's experiment with Louisiana tobacco met with mixed results. It began in the fall of 1770, when Louisiana's governor was ordered to ship all available tobacco to the tobacco monopoly in New Spain. Louisiana's tobacco production, however, did not really expand until the early 1780s, when Spain sent Intendant Martín Navarro to promote tobacco in the region.[31] According to Brian E. Coutts, "By 1782, Navarro had become convinced that tobacco offered perhaps the greatest hope for the future development of the colony."[32] The colony ultimately managed to ship over five million pounds of tobacco to New Spain's royal monopoly during the ten years between 1778 and 1788. Its production expanded so rapidly during the rest of the decade that by 1789, the importation of Louisiana tobacco was stopped until the king could be consulted. Shipments of tobacco from the Spanish colony ended altogether on 1 January 1792, leaving more than 3.7 million pounds of Louisiana tobacco in Seville's storehouses. The surplus was a result of overproduction, combined with poor packing and shipping practices that had displeased the Crown and dissatisfied Spanish and Mexican consumers. For Louisiana's planters, this spelled the end of their participation in tobacco cultivation. Those who had purchased slaves to open new tobacco land were bankrupted, leaving many New Orleans merchants who had loaned the planters money unable to collect on the loans. Even though the Crown sought tobacco again from Louisiana in 1795, not enough was available, and by the turn of the century, most of the territory's planters grew cotton.[33]

Whether they took place in Spain, Louisiana, or elsewhere, a key component in the Spanish Crown's attempts to exercise economic power over the tobacco industry and its producers was the enforcement branch of the monopoly, or *resguardo*. The *resguardo*, responsible for implementing monopoly regulations and preventing contraband in tobacco, can be dated to as early as 1764. That year, Cuba's captain general, the Conde de Ricla (Ambrosio Funes de Villalpando), assigned a unit to Regla, located just across the harbor from Havana. The corps was composed of one corporal and three guards, who were assisted by one captain and twelve mounted troops. A separate horse-mounted troop was created on 28 September 1778, comprising one primary corporal and five guards; it was known by several names, including the *resguardo a caballo*, the *resguardo de a cavallo*, and the *resguardo montado*. The official uniform consisted of a blue jacket with red lapels, blue pants, blue socks, a blue cape, and a white gaucho-style hat. Each man's armament included a wide sword worn at the waist, a pair of pistols, and a rifle. The conspicuous uniform probably served as a visual deterrent to crime, but despite their intimidating physical appearance, the *resguardo* did not have an easy task as they patrolled the area, asking those on foot what they were doing and checking suspicious spots for illegal contraband activity.[34]

The presence of the *resguardo*, be it on foot or horseback, indicated the lengths to which Spain was willing to go in order to protect the monopoly and prevent contraband. This task was far from an easy one, however (and is discussed in greater detail in Chapter 6),and the reality was that members of the tobacco monopoly police occasionally took advantage of their proximity to this valuable crop and engaged in illegal activities themselves. The mere existence of a monopoly to control the purchase of tobacco and establish fixed prices for this crop illustrates the priority Spain placed on exacting revenue from this prized commodity. Tobacco was appreciated worldwide, and the best way to ensure that the Crown could partake of this revenue stream was to control all the levels of its production, sale, and manufacture.

Tobacco also affected Cuba, Spain, and the Atlantic world through the monies spent to operate the monopoly. The most obvious expenses related to Spain's control of colonial tobacco were those tied to salaries. An examination of Cuban monopoly records indicates that in 1796, the island's tobacco monopoly workforce included fifty people who worked in general administration, accounting, notary, and treasury functions. These positions alone accounted for 35,714 *pesos* in payroll. The accounting division, the *contaduría*, employed seven people at a salary cost of 7,485 *pesos*—about 20 percent of the monopoly's budget—while four individuals worked in notary functions. Eight years later, in 1804, payroll costs for these positions had risen to over fifty-six thousand *pesos*, not including the head of the tobacco monopoly. By 1809, the monopoly expanded its paid workforce to one hundred and seventeen people, although seven positions remained unfilled, including those of the *resguardo*'s commander and lieutenant. Immediately before the end of the monopoly, the *estanco* employed eighty-one people throughout the island.[35]

Salaries for Seville's tobacco factory similarly demonstrate the financial expense Spain undertook in order to control its colonies' tobacco industry. A sampling of extant accounting books from Seville's tobacco factory reveals that salaries at the facility increased over time by slightly more than 61 percent in just fifty-five years, while the number of employees rose just over 25 percent during this period. Sal-

TABLE 4.1: Employment in Seville's Tobacco Factory

YEAR	NUMBER OF EMPLOYEES	INCREASE OR DECREASE (EMPLOYEES)	SALARIES (IN REALES DE VELLON PER YEAR)	% INCREASE OR DECREASE (SALARIES)
1762	83	—	386,836	—
1814	97*	+ 16.867%	540,000	+ 39.594%
1816	104*	+ 7.216%	618,168	+ 14.476%
1817	(unknown)		623,359	+ 0.84%

*The figure in 1814 includes ninety employees and seven retirees and is calculated based on monthly salary data. The figure for 1816 includes ninety-six employees, four retirees, and four other individuals (including widows), and is calculated based on monthly salary data.

Sources: Francisco Rodríquez Vizcayo, "Certifico que los dependientes que sirven en estas Rs Fábricas con sus sueldos anuales, graduaciones y destinos son los siguentes. . . . ," 23 March 1762, Seville, leg. 2016, AFTS; Francisco del Nero, "Borrador de la Nomina de sueldos del mes de febrero de 1814," 26 February 1814, Seville, leg. 2123, AFTS; "Importe de 12 Nominas de Sueldos Correspondentes al presente año de 1817," n.d., leg. 2124, AFTS.

aries increased from a total of 386,836 *reales* (approximately $77,367 at the time) in 1762 to nearly 625,000 *reales* (about $125,000) in 1817, the last year of Cuba's tobacco monopoly (see Table 4.1). The 1814 payroll for ninety factory employees and seven retirees for just one month totaled 45,000 *reales* ($9,000). The superintendent of the Seville factory, don José de Espinosa y Brun, earned just over 3,333 *reales* per month, for a total of 40,000 *reales* ($8,000) annually.[36]

Tobacco production varied significantly throughout Spanish America, yet the bureaucracy associated with the tobacco monopoly remained fairly unchanged. As was the case in Cuba and Spain, salaries constituted a sizable portion of each country's tobacco monopoly's budget. Wages in the Factoría General de Santiago de Guatemala during the mid-eighteenth century added up to at least 8,400 *pesos* annually, whereas Venezuela's tobacco monopoly paid out nearly 39,000 *pesos* in employee salaries in 1783. As of 1788, the majority of Mexico's *estanco* employees received lower levels of remuneration compared to monopoly employees in other locations, but the general directors and those holding the combined office of accountant and director could each expect to earn six thousand *pesos* per year, while *factores* (agents) received two thousand *pesos* annually.[37]

The funds paid to the farmers themselves demonstrate the economic power of tobacco during this time. As early as the 1720s, a portion of the *situado* from New Spain to Havana was dedicated to paying tobacco farmers. Initially, the figure totaled two hundred thousand *pesos* annually, but by the mid-1740s, the amount dedicated to compensating Cuba's *vegueros* for their product had risen to five hundred thousand *pesos* per year. Havana's military garrison in the 1740s, by contrast, received only four hundred thousand *pesos* per year from the New Spain *situado*. Between 1751 and 1755, *arriero* Pedro Hernández and Lieutenant (junior grade) don Manuel de Quintana regularly delivered funds from the island's tobacco monopoly headquarters in Havana to eastern Cuba to pay for the area's tobacco harvest. Cuba's captain general, Francisco Cagigal de la Vega, informed Santiago de Cuba's governor on 30 April 1751 that Hernández and Quintana would be sent out that day with 40,480 *pesos* 2½ *reales* to be used for purchasing the region's annual tobacco crop. This order was repeated in 1754, but these were not the only times that the pair would be asked to deliver money to the local eastern tobacco monopoly's coffers. Hernández and Quintana again left on the morning of 20 July 1755 to deliver 36,399 *pesos* in *plata sencilla* (single silver) to Santiago de Cuba, and, if needed, to Bayamo as well. Hernández and Quintana were responsible for transporting nearly 140,000 *pesos* in all during this four-year period to Santiago de Cuba's local monopoly.[38]

How was Spain able fiscally to support these expenses associated with the tobacco monopoly? The Crown saw a return of 500 to 600 percent on their investment in Cuba's tobacco monopoly in just twenty years, from 1740 to 1760. During that time, the island's tobacco *estanco* brought in between six and nine million *pesos* a year, or "about one-fifth of the entire royal income." The average annual output of Mexican mines during the early 1720s by comparison was just over ten million *pesos*.[39] The impact of tobacco profits on Spain's ability to govern and protect its strategically located colony of Cuba cannot be overstated. The profit from Cuban tobacco alone was more than capable of funding costs associated with governing and defending the island. From 1763 to 1803, a total of 2,494,120 *pesos* representing the value of unprocessed and ground tobacco sold for public consumption entered the island's Real Factoría coffers. The value of tobacco was even greater for Spain as a whole, and remained relatively constant over time. Jacques A. Barbier and Herbert S. Klein note that Spain's tobacco monopoly revenue between 1784 and 1807 averaged nearly 9.3 million *reales de vellón* annually, making the yearly salaries of the Seville factory a mere 7 percent of the average annual revenue intake.[40] Yet, it was not just in Cuba that tobacco helped to finance the colony and the Crown's expenses. Monies from tobacco *estancos* elsewhere in the Spanish Empire occasionally were used to pay for needed local projects. The governor of the Philippine Islands, Rafael María de Aguilar, used funds from its own tobacco monopoly to pay for the Philippines' increased defense costs in 1795 and 1796.[41]

For Cuba, Spain, and the Atlantic world as a whole during this period, tobacco played an important economic role in colonial society. Spain was willing to undertake the expense of operating a monopoly on tobacco, yet it came at a cost, both social and financial. Spain, in its creation and administration of the tobacco monopoly, sought to use the power of this agricultural product to command an important segment of the colonial Cuban economy. The three rebellions that occurred when the *estanco* was first established were a response to this economic control. The *vegueros* resented state management of their industry. Spain's plan to purchase Cuban tobacco at reduced prices and sell it on the open market at inflated levels angered those who made their living from this plant. Those affected by these new measures banded together to protest the economic impact on their daily lives. These men were willing to defend with their lives their right to sell their product without restrictions. Forcing out the highest representative of the colonial Spanish government on the island, the *vegueros* demonstrated in the most visible way possible their displeasure with the new economic control the Bourbon monarchy imposed on their lives. Continued resistance in the form of armed rebellions through the next several years reinforced the tobacco farmers' commitment to maintaining as much, if not total, economic control over their endeavors as possible. When it became clear that Spain would not relent in its decision to exert total authority over one of the island's best-known agricultural products, farmers and others resorted to illegal trade in tobacco. Contraband, addressed in detail in Chapter 6, offered them a form of subtle resistance. In evaluating the performance of the Cuban tobacco monopoly, it is necessary to define the factors that determine success or failure. If the ability of the monopoly to generate revenue for the Spanish Crown is used as the sole measure of its accomplishments, it can be viewed as an overwhelming success.

5 "Apply all of your attention to the laws"

EFFORTS TO REGULATE ONE of colonial Cuba's most recognizable products contained elements of economic control over the island's population; however, other methods of control were also at play. An essential component involved the creation of a vast bureaucratic system to control and regulate the tobacco monopoly. Spain held firm control over nearly every aspect of its operation, as it did with other colonial institutions, and staffed the organization with those from the peninsula. The initial response to the monopoly was dramatic, with several periods of armed rebellion following a brief attempt to negotiate with the metropole for removal of the *estanco*. Spain struggled with compliance issues, predominantly in the area of contraband. However, the economic measures proved so successful that by the end of the eighteenth century virtually every Spanish colonial entity capable of growing tobacco had instituted a tobacco monopoly.

Spain utilized the tobacco monopoly in order to exert economic control over anyone associated with the tobacco industry, ranging from growers to sellers. These limitations included on opening new land to tobacco cultivation, periodic inspection of tobacco to determine how much should be presented to the monopoly after harvest, and other regulations placed on Cuba's residents within the tobacco industry. While many of the controls aimed at tobacco sought to address profit and contraband, the tobacco monopoly also contained elements that contributed to the social control of those involved with tobacco as well. Through the creation of a vast system of bureaucracy that monitored every aspect of tobacco's cultivation and production, as well as those who engaged in it, Spain was able to assert social control over these individuals.

Although the creation of a monopoly on tobacco proved a logical method of regulating the industry, those who grew the plant were not as pleased. *Vegueros* initially attempted to reason with local authorities, hoping to remove or alter the conditions of the *estanco*. Despite several attempts, they were unsuccessful in their peaceful petitions. The third and final *veguero* insurrection ended brutally. As noted in Chapter 4, the farmers were split along socioeconomic lines regarding this last rebellion, which had not been the case during earlier revolts. Spanish authorities captured eleven tobacco growers and executed them. Spain's firm response, coupled with the harsh visual warning of the men's quartered bodies hanging from trees, demonstrated how seriously Cuban governmental officials took the new tobacco monopoly. While the *vegueros* had succeeded in 1717 in driving one captain general back to Spain, the prompt reassignment of another captain general in his place confirmed that Spain would not back down on this regulation.

While contraband and other illicit activities limited economic compliance with the *estanco*, the monopoly's existence and functions led to a level of social control on the island. The monopoly's most conspicuous responsibilities included keeping track of the tobacco produced and ensuring that farmers delivered all the tobacco designated for the monopoly. Traveling to the distant *vegas*, inspecting the properties, and interviewing the tobacco farmers allowed the *estanco*'s representative to observe the activities of those engaged in the cultivation and production of tobacco. Local officials, either from the monopoly itself (such as the tobacco monopoly agent) or the *capitán del partido*, were able to monitor a wide range of Cuba's rural population through this process. The island's tobacco industry included a vast array of those living outside of the urban areas, from the wealthy to the poor and the free and the enslaved, as well as Europeans and people of color. These individuals would have lived their lives in relative anonymity (at least where those in the colonial government were concerned) without this close supervision. Serious economic changes to their way of life, in the form of governmental monopolies, may have prompted them to resist through passive or violent means, as they initially did in the early days of the *estanco*. The results of regular supervision were reported to those at the highest levels in the island's colonial government, and possibly played a role in preventing social disorder and unrest in this important rural community. A letter from one of Cuba's captains general, the Marqués de la Torre, provides an indication of the importance of gathering information needed for a *tazmia*. José Claro found out in 1773 that this particular charge was not a job to be taken lightly. Commissioning Claro to make a *visita* (inspection) of several *partidos*, de la Torre gave Claro permission to compile data for a *tazmia*, and reminded him that the tobacco industry operated under "particular rules."[1]

Tobacco farms frequently were located in isolated areas, even those that were geographically relatively near the island's capital. One such area was located west of Havana, where rolling hills and plains rise up toward Cuba's northern coast to become the Sierra de los Órganos and Sierra del Rosario Mountains. The area's red to gray-brown soil provided good drainage, in contrast to the clay found throughout the sugar-growing districts, and made the land ideal for tobacco cultivation. Within this region, known as Nueva Filipinas—today's province of Pinar del Río—was the famed Vuelta Abajo, an area approximately eighty-one miles long by twenty-one miles wide. It was bordered by Consolación del Sur, Mantua, the mountains to the north, and a line of palm trees that ran parallel to the southern coast.[2] Countless nineteenth-century foreign visitors cited the Vuelta Abajo as containing the best tobacco in Cuba, with one even describing the area as "the promised land of the small planters of Kentucky and Virginia."[3]

A desire to exert greater control over the region's tobacco farming and its residents, particularly in circumventing contraband in tobacco, led to Pinar del Río's creation as a separate jurisdiction. Pinar del Río was very much isolated, despite the presence of an increasingly vibrant tobacco industry. Travel to the region involved a grueling trip along primitive paths and over many streams in order to negotiate the sixty leagues from Havana to the Cuyaguateje River, which traversed the area.[4] A government official in the late eighteenth century described the region as being without "an established town whatsoever, and all this vast terrain is reduced to haciendas, estancias and *ranchos* [ranches], with a few *casas de paja* [palm-thatched houses] built near the parochial churches for each district."[5] The area remained largely outside of the island's political infrastructure. Without effective local governance, it became lawless and disorganized.

Illegal activity in Cuba's tobacco industry was a particular concern, as evidenced in a series of letters from local officials to the Marqués de la Torre. Tobacco monopoly *visitador* (inspector) Francisco Jacott wrote from San Juan y Martínez in 1773 that in less than a month on the job, he had discovered tobacco hidden in "outdoor kitchens, small store rooms, chicken coops, and in some mountain-side ranches."[6] He recovered some tobacco after successfully obtaining confessions from several *vegueros*. In the presence of local *capitán del partido* Nicolás Hidalgo Gato, Lieutenant (junior grade) Vicente Blanco, and a captain in the reformed militia, Juan de Mesa, Jacott informed the local tobacco farmers that they had twenty-four hours to turn in their tobacco. Houses and estates thereafter would be searched and any remaining tobacco would be confiscated. He repeated the order to the *vegueros*, he stated in his 2 September letter to Cuba's highest government officer, to ensure that no one could claim ignorance.[7] Jacott's directive proved effective, so much so that just a few days after he issued the order, he complained that not a single mule pack driver could be found to transport the tobacco collected to the tobacco monopoly warehouse in Havana. Many had been available earlier when contrabanders frequented the coast, he said. Jacott requested three hundreds animals for the *partido* of Consolación, where they would be dispatched to the areas most in need of them to carry tobacco to the capital.[8]

Within two years, the situation in Pinar del Río had grown so severe that Cuba's captain general, the Marqués de la Torre, openly complained about it in a letter to Julián de Arriaga, Spain's minister of the Indies. De la Torre lamented, "Each day, the clamors of the disorder, injustice, and vexations of those inhabitants reciprocally caused by them, and suffered by the constables . . . reach my ears." These concerns, combined with a desire to stem the tide of illegal commerce between Cuban and foreign ships and the area's residents, led the captain general to suggest that the best way to govern the region was to formally found a town. On 23 July 1774, de la Torre proposed the creation of a new western administrative jurisdiction. Separated from Havana, the area would contain six *partidos* between the Palacios River and the Cape of San Antonio and be placed under the control of a *teniente gobernador* (lieutenant governor). This official would govern it in a manner similar to the system already existing in Trinidad, Puerto Príncipe, and other locations on the island.[9]

On the surface, this account appears to address economic concerns about the lack of compliance with monopoly regulations. A subplot to this story, however, is the element of social control. Diverting funds away from the official monopoly and to contrabanders was unacceptable, but many of de la Torre's complaints stem from Havana's inability to control the residents of this area. With additional colonial government in place, the region's relative isolation would no longer pose as significant a threat to government control. What was needed, de la Torre felt, was a strong hand locally to monitor the population and ensure that it remained compliant and neutralized. This is clearly indicated in the captain general's choice for the new lieutenant governor of the area: Antonio Fernández, a captain in the Escuadrón de Dragones de America (Dragoons Regiment of America), had served for nearly thirty-five years in the military, with a distinguished record that included combat in Europe and North America. A devoted soldier descended from a military family, the sixty-plus-year-old Fernández had been bred for loyalty to the Spanish crown, and he had embraced the "honorific career of the armed services since his infancy."[10] Fernández had entered the army in his late twenties, and served as a member of the Regimento Infanteria de Africa (Infantry Regiment of Africa). He had seen action in numerous battles in the Italian theater during his career. They had included the defense of the town of Chusanico, an attack on the trenches of Montalban, and the sieges of Dumon, Tortona, and Valencia del Po. During the battle of Plasencia, he had defended the flag he carried, never losing possession of it, even when enemy cavalry trampled him. His tenure in the Americas also featured meritorious military service and more brushes with history. In 1762, he had become a cavalry lieutenant at the Saint Augustine, Florida, garrison, where he had engaged in numerous encounters with the local Native Americans. When the territory was ceded back to Britain as part of the 1763 Treaty of Paris, Fernández had been forced to leave behind two horses and several houses belonging to his wife, Victorina Guillén.[11]

Spain dispatched Fernández to western Cuba by early 1775 to begin serving at his new post. In an instruction dated 16 January, de la Torre laid out Fernández's first duties. Fernández was to establish the limits of the new territory and to inform the local *capitanes del partido* that they were subordinate to him in political and military matters. The captain general also instructed Fernández to restore law and order, emphasizing the importance of "pursu[ing] with firmness the evildoers" and "free[ing] your jurisdiction of the wickedness, risks and fears" disturbing the public order. "Apply all of your attention to the laws," de la Torre ordered. Deserters from the veteran troops, fugitive criminals, vagrants, and other lawbreakers were specifically mentioned as deserving targets of the lieutenant governor's attention. Filipinas's strategic location also meant that Fernández was on the front line against pirates; he was ordered to alert the captain general to any ships that passed through the Cabo Corrientes at the southernmost tip of the new jurisdiction.[12]

All of Filipinas's residents were to be monitored, yet another vital component of Fernández's responsibilities was to serve as district chief of the Intendencia General of the Real Hacienda (General Intendancy of the Royal Treasury), a task assigned to lieutenant governors elsewhere on the island. This role placed Fernández in charge of all matters related to the tobacco monopoly, including the responsibility of ensuring that Havana's directives were followed. The excellent quality of the area's tobacco meant that preventing fraud, contraband, and the illicit sale of tobacco was particularly important, de la Torre noted. The chief's other duties including making and receiving payments for tobacco produced for the Real Factoría de Tabacos, and sending all the tobacco obtained from the local *vegueros* to Havana.[13]

In hindsight, the reasons behind de la Torre's choice for this exceedingly important post become even clearer. Fernández's devotion to the Spanish Crown was unshakable, and it did not end with his term as lieutenant governor of Filipinas. The dedicated soldier left the island to defend Spain during the war with Great Britain (1779–1783). Fighting under the command of Field Marshal Bernardo de Gálvez, Fernández joined in the attack on Mobile in the spring of 1780, leading twenty-five dragoons and thirty free blacks through enemy territory as they traveled from New Orleans to the Bay of Olives, where he met up with members of Gálvez's army. Fernández also participated in the siege of Pensacola, which spelled the end of British control of West Florida. At the time of his death, around 1800, Fernández had served Spain's monarchs for over fifty-two years in a military capacity, including as commander of Havana's San Carlos de la Cabaña fortress.[14]

The appointment of a loyal and dedicated officer to oversee this remote area was essential. Choosing the right location for the new *pueblo* (town), which would be named Filipina, would also be important to the success of Spain's attempt to control the region's population. De la Torre advised his lieutenant to pick a location with healthy winds and access to raw materials for construction. Ideally, it would also be along the banks of the Cuyaguateje River, which had been marked as the center of the new jurisdictional area, but Fernández could choose the locale he deemed most appropriate. Engineer Francisco Gelabert, dispatched to the area to survey the territory and assist in the selection of the final site, reported that while the current location of an existing *villa* was "the most worthy that I found in all of that territory . . . it does not appear to me to be appropriate for founding the new town." Noting that the new town was "principally for the development of tobacco," Gelabert concluded that it would be best if the settlement was located close to the tobacco farms. Land located along the *Carrera de los Vegueros* (Road of the Tobacco Farmers) was on higher ground and offered ample access to good soil for bricks and a forest that could provide lumber. Unfortunately, however, it also was nearly half a league (approximately 1.5 miles) from the Cuyaguateje River, where the area's *vegas* were located. The *Paso de los Ciegos* (Passage of the Blind) along the Cuyaguateje River was situated among the existing twenty-four *vegas*, which would make it easier for the *teniente gobernador* to oversee the tobacco farmers' activities. Additionally, the location had abundant access to potable water, construction materials, and healthy winds. Another forty-eight or perhaps one hundred *vegas* could be opened in the area, Gelabert commented. Unfortunately, the land available for the new *pueblo* in this location was limited and, as a result, the size of the town would be restricted to only two hundred *solares* (household lots). No one will move to this location, Gelabert stated, "whose interest is not founded in tobacco farming, as it is the only enterprise that will provide a living for them, since that land according to practice and time" had demonstrated its suitability only for the cultivation of tobacco.[15] Thus, if the Crown wished to monitor the activities of the new town, it needed to be located in close proximity to the *vegas*.

The tobacco monopoly's administrators supplemented the presence of local colonial officials and helped to maintain law and order. The *estanco*'s bureaucratic headquarters were located in Havana, and (as discussed in Chapter 4), administrators at the *factorías* enforced monopoly regulations at the local level. In 1761, there were ten *factorías* throughout the island, and there would be eleven by 1796. Oriente Province possessed three *factorías* (Santiago de Cuba, Mayarí, and Bayamo), while the agencies in the central part of the island were located at Holgüin, Puerto Príncipe, Trinidad, Cayo, Ságua, and Matanzas. Nueva Filipinas, the province created in 1774 and home to the island's preeminent tobacco-growing Vuelta Abajo region, contained two administrative districts, one located at Guane del Norte (also known as Mantua) and the other at Guane del Sur. By the final year of the monopoly, 1817, twenty-five *factorías* in all oversaw operations at the local level. Most of these newly created administrative districts were in located in the eastern or central part of the island. Baracóa, which had lost its *factoría* between 1761 and 1796, again was made an administrative district, along with the new locations of Guantánamo, Giguaní, Manzanillo, Tunas, Moron, Jágua, Ságua Chica, and Villa Clara.[16]

The head of each district was a monopoly representative known as a *factor*. Responsible for day-to-day administration, the factor was a key part in the success of the monopoly at the local level. The position, however, was an unpaid one, and those holding the post were prohibited from personally owning a *vega* or participating in tobacco cultivation in any form, in order to ensure their objectivity as well as their compliance with colonial policies. Authorities reminded the factor that he was to treat the *vegueros* in the "most benign and smooth manner." The factor was expected to visit each of the *vegueros* on their farms and to compile the annual *tazmias*, noting the growers' names and how many plants each *vega* was capable of holding. He was also the local expert on tobacco cultivation practices and he was to instruct the farmers in the proper, approved methods. This included ensuring that no transplanting took place before 15 October, since early plantings resulted in poor-quality tobacco, monopoly officials warned.[17] Puerto Príncipe's local monopoly representative during the late 1760s, Joaquín de Sola, acknowledged the seriousness of his position and assured his superiors that he would not disappoint them.[18] During the same period, Fernando Muñoz served as the factor for the area around Trinidad, which was closer to Havana and along the southern coast. He complained to Cuba's captain general in July 1768 that he was trying to prevent the hiding of or illicit trade in tobacco, but that it was difficult. Nevertheless, he remained at his post despite his frustration, faithfully serving the tobacco monopoly as its local representative until 1783 or perhaps later.[19] By the turn of the nineteenth century, at least three men had served as the factor for Mayarí in eastern Cuba. Temporary factor don Juan Francisco Girón held the post in 1799; it had been occupied by Luis de Echavarría during the mid-1760s, and was later held by don Miguel de Villaparo.[20]

Another monopoly employee who helped to maintain social order and compliance with the tobacco monopoly was the *guarda-almacen* (warehouse guard). A royal instruction from 1796 states that the two general *guarda-almacen* assigned to the monopoly in Havana should be men of great "fidelity, zealousness and activity" to ensure that the "service of the King does not suffer the least harm." The guard's duties included taking a general inventory of all the implements, farms, and other property belonging to the monopoly, both in Havana and throughout the island. The *oficial de libros* (book official) assisted him as he noted the amount of tobacco from each district that entered the warehouses under his control, as well as the name of the *veguero* who had sent it, all of which was recorded in another master book. The information from the general and tobacco inventories were to be turned over to the monopoly's Treasury Department.[21]

Although the salary of the *guarda-almacen* in the Surgidero de Batabanó was among the lowest among all the monopoly positions, paying just four hundred *pesos* annually as late as 1796, the job was an extremely vital one. The guard was responsible for duties associated with receiving all the shipments of tobacco that arrived from the southern part of the island. He was ordered to count and weigh the *tercios* (bundles) brought by the ship captains, to ensure that they matched the quantities reportedly sent. If they did not, he was to alert the monopoly's general administration. The warehouse's slaves were to undo all the bundles, looking for any defective leaves and removing them so that only uncontaminated bundles remained. Once this inspection was completed, the tobacco was to be sent immediately by land to the main monopoly building in Havana, as was done with all the other tobacco received by the port. This post was so important to the success of the island's tobacco monopoly that when the monopoly's board decided on 15 February 1771 that don Pedro Sanchez was to replace the deceased don José de Pedraza as Batabanó's warehouse guard, even the king was notified.[22]

One level of social control took place in the rural hinterland, where officials monitored the tobacco crop from the field to the barn and on to the factory itself. However, the Cuban tobacco monopoly was able to assert authority not only over the tobacco industry but across the entire island because of the sheer number of bureaucrats present, particularly in the urban centers. The tobacco monopoly was a sizable enterprise during the late eighteenth and early nineteenth centuries. The bureaucratic structure and operational system of the *estanco* developed and tested in Cuba was eventually copied in other Spanish colonies, despite the shortcomings it demonstrated over the years.[23]

Other tobacco monopolies used similar methods to control the social and economic activities of their populations during this same period. After 1775, Brazil's farmers were required to register with the local town councils if they were producing tobacco for export. The information provided to authorities included the farmer's "name, civil status and residence, the location of the fields, and the method of fertilization." Officials assigned the tobacco grower an identification number, and required the farmer to mark all tobacco rolls with this number. Brazilian tobacco officials hoped to prosecute those responsible for contraband tobacco by identifying its origin, and thereby to cut down on illegal activities.[24] This measure was clearly designed to encourage economic compliance with official colonial regulations, but like Cuba's efforts, it also revealed a desire for social control. Through this additional layer of supervision and monitoring, Brazil's authorities could assert social control over their rural populations.

Making the Cuban tobacco monopoly even more effective was the involvement of virtually every branch of the island's colonial government in one form or another. Their participation was not limited to occasions for oversight or even to times of emergency; government officials participated in the monopoly's daily activities. At the top of the monopoly's bureaucratic ladder was the highest-ranking official in Cuba, the captain general. Described as the *juez protector* (protection officer) in an eighteenth-century document outlining the various duties of each branch of monopoly administration, the captain general was charged with furthering the island's tobacco industry. His responsibilities included ensuring that the *vegueros* turned in all tobacco over *medio pie arriba* (six inches high) and that no tobacco was removed from the island for foreign export. Next came the *junta de factoría* (factory or monopoly board), which was immediately in charge of the island's tobacco sector and directed to encourage excellence in the quality of the tobacco produced by the *vegueros*. The *junta* also had the task of sending to Spain all the leaf and ground tobacco requested by Seville, as well as distributing tobacco to other parts of Spanish America, including New Spain and Peru.[25]

Another colonial official was Cuba's intendant, a position created in 1764. This individual was responsible for oversight of the two essential departments of treasury and war, as well as assisting in the monopoly's administration. The intendant had almost unlimited powers over the military, although his role was somewhat limited in relation to the treasury, where he dealt primarily with financial concerns. "The collection, safeguarding, auditing and expenditure of revenues and funds," increasing Cuba's revenue, and "preventing and punishing smuggling" all garnered his attention, however.[26] Cuba's first intendant, Miguel de Altarriba, arrived in Havana in early 1765 and served in this post until his health forced him to returned to the peninsula in 1773. Nicolás José Rapún, a former commissioner of the navy, was sent to Cuba to replace the ailing Altarriba. In early March 1776, the Marqués de la Torre informed Minister of Finance Miguel de Múzquiz that Rapún had died in his sleep.[27] Juan Ignacio de Urriza, who had also served as navy commissioner, succeeded him.[28]

As the third man to hold this position, Urriza would serve as Cuba's intendant for over a decade, from 1776 to 1787. The transition was not an entirely smooth one, with two members of Cuba's colonial administration vying for the home Rapún had previously occupied. The now-vacant rental property was owned by Pedro José Calvo de la Puerta y Arango, the Conde de Buena Vista and one of the first nine people in Cuba to hold titles of Castile. Both the new intendant and José Antonio de Armona, the first administrator of the mail system, claimed that it should be theirs. Urriza contended that he should be allowed to live in the house since the one where he lived was too small for the position he now held. "You know very well that the one that I currently occupy is so small that it barely holds one small family in its tightness, without giving space for the legs of the secretary, *audiencia* room, and other offices," he asserted to de la Torre.[29]

Urriza, unlike some other colonial officials, displayed the same passion for his job as he had regarding his housing arrangements. In particular, he took his order to prevent contraband very seriously. He was a strong opponent of such illegal trade, and used "the issue to destroy other public servants less dedicated to controlling this problem." He was equally concerned with demonstrating his own successes in the war against contraband, writing to Diego José Navarro (who had replaced de la Torre as Cuba's captain general) at least seventeen times over a four-month period from August to November 1779 to inform him that tobacco intended for unauthorized sale or export had been recovered.[30]

Another notable man to hold the job of intendant was Rafael Gómez Robaud, a well-connected member of Spain's government and a close friend of Joaquín Beltrán de Santa Cruz y Cárdenas, who held the titles of the first Conde de Santa Cruz de Mopox and third Conde de San Juan de Jaruco.[31] At the end of 1802, Minister of Finance Miguel Cayetano Soler personally selected Gómez Robaud to serve as Cuba's newest intendant. Gómez Robaud acknowledged in a report written after the *estanco* was abolished that he had been a poor candidate for the job as it pertained to his responsibilities regarding tobacco. "Of that income and its branch [tobacco], I did not have the least information or idea," he wrote. He confessed that he had turned to volume four of a six-volume universal dictionary written by F. Panchet in 1801 in order to learn more about tobacco, one of the most vital components of his new assignment. He also spent nearly a year collecting information and visiting some of the best French snuff factories in Brittany and other locations, which improved his knowledge of the topic, and in mid-1804, Gómez Robaud arrived in Havana, ready to begin the task set before him.[32]

Gómez Robaud was removed from his post as intendant in 1808, after another tobacco monopoly employee accused him of corruption. Nevertheless, Gómez Robaud continued on as the tobacco monopoly's superintendent. This position, like those above and below it, required full compliance with all *estanco* provisions. In response to a vacancy, the superintendent was to request from the top officer of that department the name of the person who would be the best replacement for the position, and to offer that his name to the monopoly board. As the overseer of the monopoly, the superintendent was to forward the accounting books to Spain each month, as well as shipping out the tobacco requested by other parts of Spanish America. Most importantly, the superintendent was the only official who could assign funds from the monopoly's treasury to cover payment of the tobacco farmers, employee salaries, and other such ordinary expenses.[33]

Cuba's tobacco monopoly provided many opportunities for individuals interested in moving up the colonial bureaucratic ladder, as evidenced by the career of Braulio de Vivanco. When the monopoly ended in 1817, the married, forty-nine-year-old de Vivanco was a high-ranking tobacco monopoly bureaucrat. De Vivanco's career with the monopoly began in early 1796 when he was hired as a junior clerk in the monopoly's treasury department, with an annual salary of five hundred *pesos*. He worked in this position for seven years and eighteen days, and then rose through the ranks of the monopoly's bureaucracy. He served as head clerk for eleven and a half months, and then advanced to interim treasurer, a position he held for four years, seven months, and twenty-eight days. On 1 May 1809, the position was made permanent, and he continued in the post of general treasurer until 1 January 1813, when he became the principal administrator of the tobacco monopoly. He advanced to his highest position, that of general director, in 1817, an appointment with an annual salary of three thousand *pesos*. Assigned to this job on 9 August, he would prove to be the last to serve in this position under the monopoly. The end of the monopoly, however, did not signal the end of his association with the Cuban tobacco industry. Braulio de Vivanco still reported to Spain on issues pertaining to tobacco on the island as late as the end of 1821. His duties may have played a role in his inability to act on the 1820 authorization that granted him permission to return to Spain to attend to his deceased mother-in-law's affairs.[34] More importantly, however, Vivanco's career demonstrates the care with which Spain sought to groom those responsible for its tobacco industry. It was not an enterprise that could afford repeated employee turnover. Through the continuance of those who knew the industry well, Spain ensured that the bureaucrats administering the tobacco monopoly were not only loyal, but also well trained in the responsibilities necessary for the *estanco*'s success.

Yet, Spain's ability to monitor and control the island's tobacco growers and industry was not limited to the job responsibilities of those associated with the *estanco*. Hailing from the Basque country in the north of Spain—more specifically, Vizcaya—Braulio de Vivanco came to Cuba and became part of the island's bureaucratic infrastructure. Antonio Aguilar de Cela, the inspector who had discovered the contaminated tobacco outside of the Real Factoría's warehouse in Havana (see Chapter 1), likewise was a *peninsular* (Spanish-born individual) from Seville.[35] These Spanish-born men, and others like them, constituted the backbone of the island's tobacco monopoly, ensuring that the metropole's orders were obeyed and helping to keep Cuba's rural population in check through the mechanisms described above.

Scholars have long acknowledged the importance of a Spanish military presence on the island, but they have yet to extend this attention to those who worked for the *estanco*. As noted in Chapter 3, much of Cuba's population growth during the late 1700s was due to Iberian peninsular immigration, particularly that of soldiers. This strong military presence was the result of a series of historical events culminating with the disastrous fall of Havana to the British during the Seven Years' War (1756–1763). This occurred despite the island's formidable colonial defenses, which included the Castillo del Morro. Once the city returned to Spanish control in 1763, emphasis was placed on restructuring and reinforcing the island's military and defenses. The two men entrusted with this task were Ambrosio Funes de Villalpando and Alejandro O'Reilly. Funes de Villalpando, the Conde de Ricla, was named as Cuba's new governor and captain general, and Field Marshal O'Reilly was appointed as the sub-inspector general of the militia and regular army. Threats from foreign enemies meant that military troops had long been stationed on the island, and in the wake of the Seven Years' War, de Ricla and O'Reilly further militarized Cuba. Over 11,750 regular troops were stationed in Havana alone by 1780, making "between one in two and one in three adult white males walking Havana's streets . . . a peninsular soldier."[36]

The presence of so many Spanish-born residents, as well as both military and tobacco monopoly bureaucrats as government-employed civilians, may have led to a Cuban society that considered itself more closely tied to the metropole. It was a time when Spain's Latin American colonies had begun to contemplate independence, and even to achieve it, but it came slowly for the Caribbean island, and only many decades after the mainland colonies achieved self-rule. Cuba's push for autonomy gained momentum after a slow and faltering independence movement, which gained in earnest only after the 1850s.[37]

Free individuals were not the only members of colonial Cuban society controlled by Spain's tobacco monopoly. Enslaved workers on the factory floor were also ruled by the *estanco*. A 1753 list of slaves owned by the Real Compañía de Comercio notes that nearly sixty male slaves were working in some capacity throughout the island for the royal tobacco monopoly. Just over two dozen male slaves between the ages of eighteen and sixty labored in Havana's Real Factoría de Tabacos. Nearly all (twenty-four out of twenty-six) were African-born, with those described as Carabali in the majority (eleven), followed by the Congo (six) and the Bambara (three). African slaves also worked in the island's tobacco mills. Thirty-two male slaves—all from Africa, with Carabali (nine) again predominating, and between the ages of twenty and sixty-eight—worked in the *molinos de tabaco* (see Table 5.1).[38] Few other contemporary tallies exist to indicate how many African slaves worked in the tobacco warehouse or in other capacities for the monopoly during other periods. One contemporary observer, however, commented that the tobacco factory employed an average of 120 slaves in 1765, and even more by the end of the monopoly. A 1791 census from a neighborhood known as Jesús, María, y José just outside the *extramuro* (city walls) listed fifty-four *esclavos del rey* (royal slaves), or nearly one-third of the neighborhood's entire enslaved population, as working in the Real Factoría. Early in the next century, the fac-

TABLE 5.1: Slaves in Cuba's tobacco monopoly and industry

IN THE REAL FACTORÍA, HAVANA

ORIGIN	NUMBER
Carabali	11
Congo	6
Bambara	3
Ganga	2
Mina	1
Mandinga	1
Chino criollo (Native-born boy)	1
Ingles (English)	1

IN THE *MOLINOS DE TABACO*

ORIGIN	NUMBER
Carabali	9
Mina	5
Bambara	5
Mandinga	4
Ganga	2
Aranda	2
Lucumi	1
Sape	1
Congo	1
Ingles (English)	1

Source: Gabriel Santos, "Lista que comprehende los negros esclavos de Rl Compania de esta Ciudad e Ysla, con destino en el Ramo de Tavacos, y distinction de sus nombres, naziones, Fihaziones, tachas y Edades, los mismos que constant del Ymbentario gral de Existencias del dia treinta y uno de octubre proximo pasado, formando con esta fha a saver," 17 February 1753, Real Factoría de Tabacos, Havana, leg. 1002, Ultramar, AGI.

tory counted nearly 170 slaves as laborers, including 42 women who worked as rollers. By 1812, an accounting of the monopoly's assets included sixty-two slaves in Havana alone.[39]

The monopoly controlled nearly every aspect of these slaves' existence, to a degree even greater than that of their free rural counterparts. As was the case with other factory employees, slaves were expected to work an additional hour before and after their usual schedule (7 a.m. until noon, and from 3 to 5 p.m.) during the periods when tobacco arrived at the Real Factoría. One intendant urged the factory slaves to say the rosary every night and "avoid friendships between members of the opposite sex," with the objective of avoiding sin. These slaves not only provided a valuable labor force to the monopoly, but also on at least one occasion prevented considerable economic loss to the Crown. On 10 February 1811, shortly after prayers, a fire broke out in the carpentry shop of Havana's Real Factoría de Tabacos building. Slaves from the Real Factoría and the nearby workers' neighborhood successfully fought the blaze and extinguished it. The result was little damage to the building or its contents.[40]

The social control exerted by Cuba's tobacco industry, however, affected even more people than just the *estanco*'s slaves and *vegueros*. For some of the island's indigent female population, particularly those residing at the Casa de Beneficencia (Charity House) in Havana, tobacco played a large role in their lives and livelihood. One of a few charitable societies located in the capital city, the Casa de Beneficencia housed some 226 people, of whom 54 were orphan girls, 105 indigent females, 54 slaves, and 14 others. In 1802, the facility received the right to open a factory to twist tobacco into cigars. One hundred and two girls between the ages of ten and fourteen, eighteen of whom were of color, worked to produce cigars. The following year saw a rise in the number of females of color working to produce tobacco for the monopoly. Between April and December 1803, the number of female slaves taught by white tobacco monopoly employees to make cigars grew from twenty-six to thirty-five. The number declined, however, as the decade wore on, and by 1810, 28 female slaves worked 290 days a year to produce cigars, with the majority averaging a combined 1,050 cigars per day and earning 444 *pesos* ¼ *reales* a year. Two *maestras* (supervisors) ensured that no abuse took place against the enslaved workers, who also were given Sundays and festive days off.[41]

A very visible component of the monopoly's social control, particularly in rural areas, was the presence of its enforcement branch. Dressed in their bright official uniforms of blue, red, and white, the *resguardo* visually conveyed the tobacco monopoly's refusal to tolerate disorder, either economic or social. If any residents needed an additional reminder, they could see the sword each member of the *resguardo* openly displayed at his waist, along with a pair of pistols and rifle.[42]

The tobacco monopoly and its bureaucrats were responsible for oversight of the island's tobacco crop, and controlled all aspects of its production. This included designating areas permitted to cultivate tobacco, and setting the prices paid for delivered tobacco. Unlike other locations in the Spanish Empire, Cuban tobacco was grown without restriction virtually everywhere on the island.[43] Spain was concerned, however, about the amount produced, and regulated this aspect of Cuba's industry. Together, the Crown and tobacco monopoly officials determined whether new land could be opened for tobacco cultivation. When production skyrocketed between 1772 and 1774, Spanish authorities took action. An astounding 765,266 *arrobas* of tobacco, more than the previous five years' total combined, entered the monopoly, causing Seville's facility to have a surplus of tobacco by 1774. Spanish officials determined that an attempt should be made to slow production, while maintaining the current number of *vegas* and ensuring that a quality product was delivered. *Vegueros* across Cuba in 1775 and 1776 were banned from creating new *vegas* as a result. The plan worked, perhaps too successfully. It would take fourteen years until the island's tobacco production would again meet or exceed the levels achieved in the early 1770s, particularly when climatic conditions failed to cooperate, with numerous storms and droughts battering the island.[44]

Faced with declining tobacco production beginning in the 1780s, the monopoly could also use its power to promote tobacco. Tobacco growers in the Partido de Batabanó were specifically encouraged to increase the number of *vegas* in the area in 1784. This was not the only location where the Crown made a concerted effort to promote tobacco, and nowhere was this better demonstrated than in the nearby town of Güines. In the late 1750s, Güines was a small population center of just forty houses surrounding the modest church of San Julián, the village's patron saint. Tobacco production in the area exploded during the 1770s, with the number of farms increasing by 111, rising from 168 to 279 *vegas* in a two-year period alone (1770 and 1771). The majority (87 percent) of the district's population of 2,340 lived on the area's tobacco fields by mid-decade. Together, these *vegas* occupied fully one-fifth of the area's land, while in contrast, the *partido*'s three *ingenios* (sugar estates) were home to less than 5 percent of the residents. Although it took more than a decade to implement, the plan to reorganize Güines as a center for tobacco cultivation that could supply the monopoly with twenty thousand pounds of leaf was a success.[45]

In addition to enacting the regulations addressing the number and location of Cuban tobacco fields and the amount of tobacco produced by its farmers, officials held strict control over the exportation of tobacco. Tobacco was to leave the island only under the auspices of the Real Factoría, a mandate imposed on Cuba as early as 1749. Captain General Francisco Cagigal de la Vega repeated this order, which was to be viewed as an "absolute prohibition," at least seven times over the course of sixteen months. Following the return of Havana in 1763, Spanish officials quickly sought to regain control of the tobacco industry. That year an edict required that the *vegueros* deliver all their tobacco to the Real Factoría and strictly prohibited the sale of tobacco to any individual or entity outside of the monopoly. The penalty for those who purchased such contraband tobacco was fifty ducats.[46] This order was repeated by top officials in Spain and disseminated by Cuba's captain general to local officials throughout the island. The frequency of these instructions indicates that the *vegueros* typically failed to comply with these orders throughout the existence of the monopoly.

While the monopoly was successful in exercising some power over the island's residents, it needed to offer fair payment to the farmers if it wanted to ensure that they brought all their crop to market. This was often easier said than done, unfortunately, making the ever-present temptation of contraband even more alluring to Cuba's tobacco farmers. The official, established prices for tobacco were supposedly based on the leaf quality, yet farmers accurately complained that different standardized prices were being applied to leaves of the same quality depending on one's location on the island.[47] Between 1765 and 1785, the monopoly paid thirty-five *reales* per *arroba* for *largo* (top-grade) leaves in Güines and Xiaraco, and thirty-three *reales* per *arroba* in Matanzas, but only twenty-four *reales* per *arroba* in the eastern part of the island known as the Tierra-adentro (the interior). The monopoly paid sixty *reales* per *arroba* for *largo* tobacco produced in 1803 and sixty-four *reales* per *arroba* in Güines, Xiaraco, and Matanzas the following year. In the Tierra-adentro, prices remained significantly lower, with *largo*-grade tobacco bringing the farmers forty-one *reales* per *arroba* in 1803 and forty-eight *reales* per *arroba* in 1804.[48]

These differences cannot be attributed strictly to the superior quality of one tobacco as compared to another. In 1802, the royal tobacco monopoly earned no less than 37 percent and as much as 58 percent net profit on Cuba's tobacco after all costs and expenses were calculated. The highest percentage of profits, 58.20 percent, was from Guane, where earnings totaled 66,672 *pesos*, while the second-largest percentage, 54.625 percent, was from the interior, where prices paid for tobacco were historically low.[49] This suggests that either the quality of tobacco from the interior was higher than the monopoly-paid prices indicated, or that the monopoly was able to sell lower-grade tobacco on the world market at over-inflated prices, resulting in a higher profit margin for the Crown.

Further strife between the government and tobacco farmers resulted from the failure to pay the *vegueros* on time for crops delivered as required to the royal monopoly (discussed in detail in Chapter 6). This was not a one-time occurrence, but a predicament that continued to create difficulties until the very end of the monopoly, particularly for farmers in the eastern part of the island.[50] Monopoly and government officials, however, were not ignorant of the effect that lack of payment had on the *vegueros*' satisfaction with the *estanco*, and consequently on their agricultural efforts. In 1805, José María de la Torre was able to encourage the tobacco farmers "with hopes of a better contract and considerations in future ones." Intendant Rafael Gómez Robaud, head of the tobacco monopoly, acknowledged in 1810 that when the monopoly paid the farmers for money owed for previous years' crops, it helped to re-stimulate the tobacco industry.[51]

Cuba's tobacco monopoly exerted a significant amount of control, both in terms of social and economic influence, over the island's tobacco farmers; however, the *vegueros* were not without recourse. They could turn to contraband, a fear constantly expressed by government and monopoly officials, or decline to pay for municipal projects as a way of expressing their displeasure. *Vegueros* in various parts of the island in 1775 and 1776 refused to pay for the construction of a bridge, the paving of Havana's city streets, construction of the Jubajay barracks, and maintenance work on the Rio Hondo barracks. They stated that the Factoría de Tabacos owed them for their crops and that they were awaiting payment from the monopoly before complying with public works duties.[52]

Cuba's tobacco monopoly necessitated the creation of a new colonial bureaucracy and the expansion of duties for offices already in existence. As a functionary of the Crown, the captain general of Cuba was expected to execute all orders from Spain, and the tobacco monopoly was simply another one of his responsibilities. Yet for those holding the office of *teniente gobernador* (lieutenant governor), tobacco was one of their primary concerns. Charged with developing the tobacco industry as much as possible, the *teniente gobernador* had to send the captain general copies of the licenses for all tobacco shipped to Havana. Cuban officials were responsible for the operation of the tobacco monopoly, including day-to-day enforcement of the regulations, but royal officials in Spain all the way up to the king himself remained aware of minute details regarding the royal tobacco monopoly's successes, and struggles. Prevention of contraband was an important concern for Spanish officials, so much so that correspondence between Minister of the Indies José de Gálvez and Havana from 1786 and 1787 reveals that Spain was interested in who would fill a vacancy within the *resguardo* unit caused by the death of one of its members. Even the details of the *resguardo*'s daily operations interested Spanish officials.[53]

Quality control was another area that deeply concerned Spain. In an 1803 royal order, King Carlos IV (reigned 1788–1808) asserted that there had been a "noticeable decline in the plantations of tobacco of the island of Cuba," and that the ground tobacco coming from Cuba was of "bad quality," even though a previous royal order in 1796 had put the monopoly on notice. The king stated that more measures needed to be taken to remedy this situation, increase the cultivation of tobacco, and preserve its excellent quality. To that end, he named Rafael Gómez Robaud as sole director of the Real Factoría de Tabacos in Cuba.[54]

Through the presence of numerous Spanish-born tobacco monopoly officials, and oversight from Spain itself, Cuba's tobacco industry and its farmers were closely monitored by the metropole. This close supervision, combined with other factors, as discussed elsewhere in this study, may have contributed to a population that was less motivated to rebel and seek independence from Spain. With a guaranteed buyer for its best-known agricultural product, Cuba maintained a level of economic stability, if not excessive profits. As a result, tobacco and Spain's *estanco* may have helped Cuba maintain its status as Spain's Ever-Faithful Isle for much longer than it may have had the *estanco* not been in place.

6 "I work with all my might to eradicate the evil"

TOBACCO CONSUMPTION was unstoppable. Although European royal houses saw the potential for great profit by cashing in on this new habit with tobacco monopolies, not even this approach could guarantee that the valued product would reach their market. Wars and other transatlantic shipping disruptions repeatedly wreaked havoc with the delivery of tobacco to waiting European consumers. Another imminent and omnipresent issue, contraband, frequently occupied Spain's tobacco monopoly officials at home and on the island. Contraband and the illegal trade in tobacco outside of the official monopoly posed an economic loss to Spain and other nations hoping to capture the plant's profit.

This illicit commerce affected Cuba, Spain, and the Atlantic world in other ways as well. Smuggling offered an alternative economic reality, one in which those participating in the illegal trade were offered a measure of independence not allowed in an otherwise restrictive, inefficient economy. For those who successfully circumvented the system, smuggling demonstrated the vulnerability of the European powers. This kind of illegal activity, particularly for those who managed to engage in this type of trade without capture, provided an avenue of commercial activity outside of the official tobacco monopoly. One could sell tobacco not only to the Spanish crown, but also to other purchasers who would gladly pay for this much-valued product.

Tobacco farmers were not a homogeneous group, and for many of them, tobacco cultivation was not their sole source of financial stability. Compliance with the monopoly therefore was not a matter of economic life or death for many of the island's *vegueros*. The absence of violent or organized protests from 1763 to 1817, however, did not mean that the Cuban *vegueros* obeyed the monopoly's regulations without hesitation. Various methods of undermining the tobacco monopoly included contraband, improper packing of one's tobacco in order to defraud the government, and other means of passive resistance. Yet, contraband also brought violence and increased tensions to the local community. Armed government patrols routinely scoured the countryside looking for those engaged in illicit behaviors, and neighbors were prompted to turn in one another in return for a portion of the goods seized.[1] Spanish regulations clearly stated the consequences for disobedience. Nevertheless, those living in the colonies were bound to adhere to these rules only as their own consciences dictated. Spanish officials could point to their continual yet futile attempts to control the local populace. As long as bureaucrats could assure the Crown that they were following Madrid's directives, however, these men remained loyal servants of the king. Some administrators took advantage of the tempting prevalence of illegal activities to avail of it themselves.

The surreptitious nature of contraband makes the study of such illicit activities difficult to substantiate or quantify, particularly since the very aim of the smuggler was to elude capture. Modern historians often must envision the magnitude of these activities and those who participated in them, since by their very nature, the historical record reflects only part of the story. The flourishing contrabander did not enter the official record, while his unsuccessful counterparts did. Official responses to such illegal trade, along with successful seizures of contraband, nevertheless can be used to determine the prevalence of this type of activity. Spanish officials struggled with this insurmountable problem. Yet, both in distant hamlets and in the island's capital, they recorded their repeated attempts to combat this most pressing issue. These numerous records allow scholars today to document what otherwise might have been silent crimes lost to history.[2]

Contraband in tobacco remained a constant and difficult problem for Spain and its Cuban tobacco monopoly. Even after the *estanco* was abolished, the Spanish officials charged with overseeing the island's economic activities continued their complaints against the practice. "The contraband that exists throughout the length of the island has always been and is very large, and in particular by foreigners; and despite the very rigid orders that have been given for this purpose, they will always be ineffective," lamented Braulio de Vivanco.[3] De Vivanco, the last man to hold the position of monopoly general director, did not overexaggerate. Cuba enjoyed a long history of participating in contraband, and the island's geography made deterring such illegal exchanges even more difficult.

Cuba's numerous ports, cays, and inlets, like the West Indies as a whole, "required an army to cover them," Captain General Juan Francisco Güemes de Horcasitas, the future Conde de Revillagigedo, acknowledged in 1741. Spain attempted to address the issue as early as 1513, by resorting to the use of two caravels to patrol Cuba's coastline. Additional *guardacostas* (coastal patrol ships) were added as the island's populated areas expanded. Havana continued to attract its share of illicit trade because of its importance as a primary port and as the Spanish treasure fleet's last stop before heading out into the Atlantic. The corsair presence was so strong along Havana's coast in 1593 that it forced the incoming Spanish governor, Juan Maldonado, to approach the city via the Old Channel and eventually travel overland from Matanzas.[4]

In the mid-sixteenth century, Spain attempted to protect the ships' valuable cargo by devising the *flota* (fleet) system. A convoy of ships left the Central and South American coasts and converged on Havana in the spring. There, they were outfitted for the long Atlantic crossing in early summer, before the peak of the hurricane season. Between two and eight armed galleons accompanied the *flota* to protect it against privateers or pirates, and, as a result, the entire fleet was rarely captured. From 1585 to 1603, at least seventy-six English expeditions intent on seizing Spanish ships headed to the Caribbean, including two led by Sir Francis Drake. One author even argues that the main purpose behind the failed settlement at Roanoke Island (located in present-day North Carolina) was to create a base from which English ships could attack the Spanish Indies and the treasure fleet. Settlements on Jamaica, Curaçao, and St. Eustatius similarly were organized in order to take advantage of illegal trade with other European powers' neighboring colonies.[5]

These measures generally worked to prevent seizure of the entire treasure fleet, but they proved particularly ineffective in curbing those determined to engage in piracy and privateering on a smaller scale. The arching string of both large and small islands that separated the Caribbean Sea from the Atlantic Ocean, known as the West Indies, had long been an area of interest. As those responsible for administering the island's tobacco monopoly during the mid-eighteenth to early-nineteenth centuries would learn, tighter regulations only prompted repeated smuggling, to which the state responded with even harsher regulations and controls. This contributed to the vicious cycle of tighter regulations and increased levels of contraband in tobacco.

Most contraband entered through the south, notably by way of Bayamo in eastern Cuba. In 1602, eight of the town's residents who had engaged in smuggling were charged with "rebellion and sentenced to death and confiscation of their property." Bayamo's *cabildo* just over one hundred years later attempted to explain to the Crown why its residents engaged in such illegal activities, arguing that these actions did not represent disloyalty to the king and his royal laws, but rather were prompted by necessity.[6] Foreign traders remained particularly active in eastern Cuba during the 1740s and 1750s. A letter dated 11 September 1747 informed the island's captain general, Francisco Cagigal de la Vega, that English ships frequented the area around Bayamo and Puerto Príncipe and introduced contraband goods. Ten years later, authorities were alerted to the arrival of two ships from Holland and Curaçao intent on engaging in illicit trade along the southern coast. Spanish officials increased troops or coastal patrols in areas known to be frequented by contrabanders, but not even the threat of the death penalty ended the illegal trade between Cubans and foreigners.[7]

One colonial Cuban governor reported grimly to Madrid during the early eighteenth century, "There are numbers of ships that pass here with no legal cargo. . . . I work with all my might to eradicate the evil."[8] Making the task all the more difficult was the lack of assistance provided by the local residents. The populace "complain[ed] bitterly" about the cost of maintaining the *guardacostas*, not because of the expense, but "simply because they hamper the illicit trade," a Cuban governor conceded in 1728.[9] Centuries of participation in illicit commerce, together with the brief period of free trade enjoyed by Havana during the British occupation (1762–1763), resulted in a population that continued to view contraband as a legitimate commercial activity. Official correspondence, which was particularly voluminous from the 1760s to the 1780s, demonstrated the continued presence of and problems associated with contraband in Cuba. The English were frequently cited, but Cuban bureaucrats also accused nationals from France, Holland, and other countries of engaging in illegal commerce.[10]

Certain commodities, especially those under governmental control, garnered special interest from local and foreign contrabanders alike. Tobacco proved to be one of the most popular goods available to smugglers throughout Spanish America.[11] "Tobacco has been one of the chief factors causing these coasts to be so much frequented by pirates, by reason of the great demand there has always been for it," a governor in Caracas complained to the Spanish Crown in 1607.[12] Spanish officials, though, were not the only ones who were struggling to control the tobacco trade within their borders with little success. Remote or less policed areas such as the Isle of Man, located in the Irish Sea, were natural centers for illegal trade in tobacco, just as Cuba was. The tiny European island's free ports provided an easy way during the seventeenth and eighteenth centuries for smugglers to elude customs, and, from there, to ship tobacco illegally to nearby England, Scotland, or Ireland. So much of England's early tobacco arrived through illicit means that by 1620, tobacco imports were required to bear a government seal. Portuguese officials similarly were concerned about the traffic in contraband tobacco throughout their colonies, as well as at home, where illicit tobacco sold openly "at prices lower than those set by the crown."[13]

Consumers worldwide particularly prized Cuban tobacco, and given the island's long history with smuggling, it is not surprising that when faced with a monopoly on their product, the island's tobacco farmers resorted to contraband. Up to 75 percent of Cuba's tobacco production was lost to illegal trade. This point was emphasized further by the routine presence of Havana leaf on the Dutch market as early as two years after Cuba's monopoly was created in 1717. Spanish authorities took measures to prevent illegal tobacco planting, smuggling, and other fraud, including an active campaign against smugglers, which contributed to the continued warfare between Spain and Great Britain (War of Jenkins' Ear [1739] and War of the Austrian Succession [1740–1748]). The government meanwhile openly acknowledged at all levels that Cuban tobacco farmers engaged in contraband. Havana regularly issued proclamations known as *bandos*, reminding the farmers that they should deliver all of their tobacco to the monopoly and to not sell any of it themselves—a clear indication that officials understood that the *veguero*'s inclination was to do otherwise. Tobacco monopoly bureaucrats particularly were infuriated by the sale of tobacco from western Cuba's Vuelta Abajo region. They complained that the farmers sold "all their tobacco" via illegal channels, leaving only "inferior" tobacco for the monopoly.[14]

Local officials comprehended the importance of preventing contraband. The lieutenant governor of Bayamo, Juan Gemir y Lleonart, was one such man. Gemir y Lleonart had established himself as a staunch enemy of illegal traders as a young cadet, and now assured Captain General Bucarely of his attention to the tobacco harvest, which he comprehended was "very important to the service of the king." Another official, Joaquín de Sola, responded similarly to memoranda from Havana. De Sola informed Bucarely that he understood that he was required to fulfill the "commission that he [Bucarely] had consigned regarding the harvest of the tobacco of this *partido*, and that how, as my duty, I am responsible for any omission that I knowingly make for the completion of said commission." The task set before local officials was formidable, noted Fernando Muñoz from Trinidad. "I am alone," he complained, a situation that made it particularly challenging to patrol the *vegas* to ensure that the farmers had not hidden any tobacco. This task was "even more difficult in the *partidos* leeward of this city, which are twenty leagues distant." Muñoz nevertheless reassured the captain general, "You can be certain that you have always procured the best from me . . . because my desire is to fulfill my superiors' orders."[15] Officials in Cuba made a point of explaining to those in Spain that their inability to fight contraband was not a result of their personal incompetence, but because of a lack of resources. If only more men, particularly those with similar desires to support the government's task, could be found to assist these local officials in their duties, perhaps contraband could be lessened or prevented altogether.

Successes, when they occurred, were met with enthusiasm at the highest levels of the Spanish colonial bureaucracy, underscoring the importance placed by Madrid and Havana on preventing the unauthorized sale of tobacco. The Consejo de Indias (Council of the Indies) was the division responsible for imperial administration. During February and March 1753, it devoted three separate sessions to recent achievements, notably the seizure of tobacco. On 17 February, the council celebrated the previous year's capture by Captain Bernardo Lorenzo Hernández of nearly two hundred pouches of ground tobacco. The tobacco, destined for the illegal trade, had been found aboard the schooner *San José y las Animas* at the port of Banes. March's council sessions were even more encouraging. This time, the Council of the Indies presented information related two more attempted cases of contraband. The council reported that another vessel named the *San José y las Animas*—this time a small schooner under the command of Marcos Ferman—had been captured in 1752. The ship, owned by a man identified in the historical record only as "Domingo" from France, had been heading toward Trinidad, and had intended to take its cargo to foreign colonies. Cuban authorities, however, had recovered forty-five sacks of tobacco along the beaches of the Río de la Chorrera. The council recorded by the end of March an even larger seizure of 210 pouches of tobacco, discovered by Captain Tiburcio Crespo in the Puerto de la Ortigosa, a location not usually used for commerce. The site was difficult to reach by land, leading officials to conclude that the tobacco had probably been brought there by sea. Three more Council of the Indies sessions were dedicated over the next decade to the topic of contraband Cuban tobacco, yet there is no evidence of illicit trade slowing down.[16]

Officials in Cuba and Spain in the ensuing years reiterated to their superiors that tobacco continued to be traded by individuals outside of the official monopoly, despite numerous orders that such activity was strictly forbidden. The situation on the island was so severe by the early 1770s that Minister of the Indies Julián de Arriaga felt obliged to remind Cuba's captain general of the standing order.[17] Three seizures during the early 1770s demonstrate the volume of this unlawful trade. A November 1772 raid on one home produced ten *tercios* (approximately one thousand pounds) of tobacco still on its stalks, while another just fifteen months later led to the confiscation of eighteen *tercios* of Vuelta Abajo tobacco found at Antonio Alfonso's *estancia* in Cruz de la Padre.[18] A local official in Guanabacoa the following year alerted the intendant general of the army about "various houses in that town" suspected of harboring illicit commerce. Eleven *tercios* of tobacco were found in Juan González's residence after further inspection. These three incidents alone resulted in the recapture of nearly four thousand pounds of contraband tobacco, or slightly more than one-third of the Guane district's entire production for that year.[19] Even seizures of this scale did little, unfortunately, to stem the flow of tobacco to sources outside of the monopoly.

Spanish and Cuban officials remained undeterred and continued to battle the illegal trade with any means available. One such method meant to dissuade others from taking similar actions was dealing harshly with smugglers. A case representative of this approach was the seizure of tins of ground tobacco and cigars illegally carried aboard a mail ship ready to set sail for Veracruz in 1769. Spanish officials captured the captain, don José Aedo, and pilot, José Antonio Vrisote. The men were taken to Havana's Castillo de la Real Fuerza, stripped of their respective jobs for life, and forbidden to work for the king in similar employment. Forced military service was a frequent penalty for those charged with fraud or contraband in tobacco, as Alejandro Mora and Lorenzo Cendra learned. Mora was sentenced to serve his time in Puerto Rico, while Cendra was sent to serve with the *batallion fijo* (fixed battalion) in Santo Domingo. Several local Cuban administrators offered their own suggestions as how the illegal trade might be prevented, such as Juan Gutierrez's proposal that a "faithful man" be used to escort ships carrying tobacco until the vessel reached deep water twelve leagues away from the Cuban coast.[20]

The emphasis shifted in the mid-1770s from fostering compliance with official orders to a more proactive stance that relied heavily on policing and seizures. The island's citizens were a valuable part of this new initiative, and informants who reported illicit trade in tobacco were rewarded with one-third of the confiscated tobacco's value. A particularly visible part of this tactic was the 1778 addition of the *resguardo montado* (horse-mounted police units) to the already-existent enforcement branch of the tobacco monopoly, the *resguardo*.[21] The success of the tobacco police in just the first year of the *resguardo montado*'s presence suggests that such an approach to contraband was effective. During this period, the number of confiscations rose dramatically, and between mid-September and mid-November 1779 alone, ten separate raids netted 157 *tercios* (about 157,000 pounds) of tobacco returned to the Real Factoría in Havana. Two original *resguardo montado* members, don Manuel Pablo Sorrivas and don Juan Morel, accounted for 40 percent of these captures.[22] This change in tactics demonstrates the prevalence of contraband on the island. Only personal profit would entice local residents to report illegal activity, officials believed. While this measure appeared to have been relatively successful, particularly if the number of captures can be used as an indicator of success, it did not end this illegal activity.

While luck, the lure of a reward, or perhaps both may have made the two particularly vigilant, another explanation for their remarkable job performance is less noble. Morel's eventual fate suggests that his early accomplishments were a result of his inside knowledge of the island's contraband tobacco trade, rather than a devotion to his duties. Sometime before September 1785, Morel, along with three other *resguardo montado* members, took part in a seizure of contraband in Guanabacoa. Their actions during this incident resulted in their being stripped of their posts and placed in prison where they remained until at least July 1787.[23]

The lure of the illegal trade in tobacco remained irresistible for many, however.

Cristóbal de Cordova risked prosecution during the mid-1780s by forging the signature of an official in western Cuba as part of an attempt to embezzle tobacco. Authorities captured de Cordova and turned him over to the island's provisional captain general, Luis de Unzaga, in Havana. The circumstances under which tobacco was seized also show the lengths to which those intent on engaging in illicit commerce would go to evade officials. Residences and farms were some of the locations where smugglers hid their product; for example, during the spring of 1800, guard Bartolomé Castellanos recovered thirty-nine *tercios* of tobacco from a home on an *ingenio* (sugar estate) owned by don José Ramos. There were far more unusual sites where authorities discovered the plant, however, such as in a hen house on an *estancia* owned by don Ramón Año y Vega. There the head of the *resguardo*, don Manuel Trensa, along with two guards, seized twelve *tercios* of San Juan y Martínez tobacco in September 1816. Monopoly officials noted that the hen house raid netted contraband valued at 101 *pesos* and one *real*, all of which Trensa returned to the Real Factoría in two wagons that he personally rented for one peso four *reales*.[24]

Those charged with tobacco smuggling in Cuba occasionally explained their actions. Tobacco farmer don Miguel Delgado of the Partido de Ursulica acknowledged that the tobacco recovered from a house located on José Ramos's *ingenio* was his. Delgado maintained, however, that Ramos had agreed to keep the 39 *tercios* (approximately 3,900 pounds) of tobacco on his sugar plantation while the farmer searched for a mule pack driver and the funds to pay for taking the tobacco to the Real Factoría. The local *diputado*, don Simon Valiente, could support this claim, Delgado argued. Another case likewise involved a tobacco farmer. This time the *veguero* was from the San Juan y Martínez district, deep in the heart of the tobacco-growing region of western Cuba. The captain of the district, don Armonio de Puentes, charged Juan de Acosta with failing to turn in twenty-five *cujes* of tobacco from the previous year's crop. The *veguero* argued that he had been unable to do so because he was "gravely ill." Eventually the complaint was lowered to ten sticks, which de Acosta claimed were for his own use.[25] Those charged with tobacco contraband in each of these instances initially had a valid reason to possess the plant—they grew it. When confronted with charges of illegal possession, however, they responded with explanations that stressed their interest in complying with proper tobacco monopoly procedures. It was only extenuating circumstances, poor health, or lack of finances that had prevented them from doing so.

Spain expected tobacco to be safe once it reached the Real Factoría and its agents, but this was not always the case. The historical record contains examples that demonstrate the temptation that tobacco posed to those willing to assume the risks of their illegal actions. In one incident, tobacco was stolen from mills in eastern Cuba and put on board the brigantine *Nuestra Señora del Amparo* in an attempted case of contraband during the late 1760s. In another case, someone took tobacco from Santiago de Cuba's royal *almacenes* (storehouses) to the wharf, and sold it for less than the official price.[26] It is difficult to tell in such cases who was responsible for taking the tobacco, particularly when the individuals were not captured. Nevertheless, speculation about their identity is possible. The primary suspects clearly are those directly associated with the monopoly itself. These people had not only access to the plant, but also the opportunity to abscond with it and to sell it outside of the *estanco* for their own profit.

Cases throughout Cuba as well as elsewhere in the Spanish Empire indicate the participation in illegal activities by those directly associated with the tobacco monopoly. In October 1768, Cuban captain general Bucarely complained that much of the tobacco carried along the route from the interior to Havana was sold before it even reached the city. Little had changed by the mid-1770s as authorities frequently seized tobacco just outside Havana's walls or within the capital itself. Night, in particular, offered the ideal cover for those attempting to bring illegal tobacco into the city. Cuban officials, however, took visible measures to secure the city and its port. Every evening at nine o'clock a cannon shot announced the closing of the walled city's five gates, and a log chain was stretched between the fortresses of El Morro on the north and La Punta on the south, effectively securing the entrance to Havana's bay and harbor. If a contrabander managed to get inside the walled city before the gates closed for the evening, however, he or she easily could elude authorities. The city's streets had no public lighting system until 1786, when the large buildings on the Plaza de Armas and street corners were illuminated. Nevertheless, Cuban officials were diligent, and their continued campaign against contraband was particularly successful in the summer of 1776. Officials recovered just over 1,400 pounds of western Cuba tobacco in three separate raids during July and August of that year. The bulk of the contraband, twelve *tercios* of tobacco from the Govea district, was found in Juan Sanchez's home on the Calle de la Muralla, just inside the city walls. The remaining tobacco, from the *partidos* of Govea and Guane, was confiscated at the *Puerta de la Tierra*, just west of the Calle de la Muralla.[27]

In his 1768 letter, Bucarely did not specifically cite mule pack drivers as responsible for theft of tobacco, but like the *vegueros*, those transporting tobacco had easy access to the plant and, if so inclined, could readily turn to illegal trade. *Arrieros* from rural settings were not guaranteed regular work and were more likely to hold another job, providing them with fewer opportunities to succumb to the temptation of contraband. Mule pack drivers near high population centers, by contrast, usually labored full time in the transportation sector. The proximity to such a valuable commodity was impossible to resist for some mule pack drivers. The nature of their job also ensured that they would rarely raise suspicion if they traveled about the countryside or city transporting tobacco or other items. Tobacco could be carried openly under the guise that it was on its way to the factory, or it could be hidden as part of a larger cargo. A local Cuban official in San Juan y Santiago noted in 1772 that several *arrieros* had failed to deliver their shipments to the Real Factoría. Several years later, on the night of 3 October 1779, the head of the *resguardo montado*, Sorrivas, apprehended mule pack driver Juan de Cristo with tobacco in the Barrio de San Luis Gonzaga, located outside of Havana's walls.[28] In another example of an *arriero* operating outside of the law, a driver named Nicolás left behind his load of tobacco as he fled the scene of a crime: he had wounded a slave assigned to work on Havana's projects, Joseph Antonio Sanchez. (Another *arriero*, Josef González from the town of Santiago, took pity on the wounded man and brought him to Havana to the royal hospital, Nuestra Señora del Pilar.)[29]

Spanish officials on the island and throughout the empire bemoaned their struggles against the seemingly ever-growing tide of illicit trade. Nonetheless, the success of contraband ultimately depended on government officials turning a blind eye to it or participating in it themselves. Illegal trade in tobacco was rampant on the island, as authorities openly acknowledged. The astonishing ability of *resguardo montado* members Sorrivas and Morel to secure so many captures raises many questions, not the least of which is the reason for their extraordinary success.

Both men were founding members of the mounted tobacco police force created in 1778. Sorrivas served as first corporal with a salary that year of 730 *pesos*, and remained at that rank until his death in September 1785. He was particularly successful in apprehending tobacco contrabanders during his tenure. Sorrivas personally was responsible for capturing at least 122 *tercios* (approximately 12,200 pounds) of tobacco and returning it to the monopoly within a two-week period from late September to early October 1779. The hiding places included, among other locations, four separate homes in the village of Guanabacoa, one belonging to *regidor* (town councilman) Agustín Hernández Pulgarón.[30] One of the first guards assigned to the mounted *resguardo*, Morel served under Sorrivas and earned a salary of 547 *pesos* in 1778. The late summer and fall of 1779 was a successful one for him, as it was for his commanding officer. In early August, Morel discovered twelve *tercios* of tobacco carried by an *arriero* between Horcón and the still-under-construction Castillo del Príncipe, a fort located west of the city, and in October, he recovered seven *tercios* of unprocessed tobacco hidden in Tomás Roque's home in the *partido* of Jesus del Monte. Several days later, in Havana's Barrio de Jesus María, Morel caught Ramón de Ochagavia unloading raw tobacco. Morel's success at capturing tobacco continued into the following month, when he seized a horse loaded with seventy-six *manojos* (bundles) of tobacco in the Partido de San Juan during the night of 6 November.[31]

The history of the mounted tobacco police during the mid-1780s in particular indicates the ongoing and viable temptation facing all members of Cuban society. By September 1785, the mounted tobacco police was in disarray, and in the fall of

1787, it effectively was dismantled, with all its living members except one in prison on serious charges. Founding *resguardo* members Juan Morel, Antonio Bonet, and Pedro Galan were arrested with relative newcomer Joaquín Cano sometime before the fall of 1785, following a confiscation in Guanabacoa. Another founding member of the *resguardo montado*, guard don Francisco Serrano, similarly fell victim to the lure of contraband. Authorities implicated Serrano in 1787 on four counts of keeping tobacco that should have been delivered to the monopoly. On 25 October, José Francisco Grillo, a twenty-five-year-old resident of Govea, gave sworn testimony before the captain of the *partido*. According to Grillo, Serrano and another individual posing as an informant had burst into Grillo's home and, without providing any explanation, seized tobacco Grillo had kept for his own use. Before he left, Serrano had demanded four *pesos* in *reales*, as well as chocolate and bread, Grillo stated.

Moreover, Grillo was not the only complainant against the guard. José Hernández Estevez, age forty-five, likewise testified that Serrano had forced his way into his residence and, under the pretense of confiscating them, taken two bundles of tobacco. In similar sworn statements, thirty-five-year-old Manuel Pita and thirty-eight-year-old Salvador Gomez maintained that Serrano had also entered their houses and removed tobacco.[32] It is possible that Serrano's position as a tobacco monopoly police member had put him at odds with certain members of the local population. Nevertheless, the cases against Serrano and other members of the *resguardo* seem to indicate that illicit activity was prevalent among these men. This may seem shocking, but as Alfonso W. Quiroz has noted, in the nineteenth century, bureaucratic corruption was not a new trend in Cuba, or even in Latin America.[33]

Members of the *resguardo* were not the only tobacco monopoly officials who were accused of misconduct. The lieutenant governor of Bayamo, Antonio Laso de la Vega, charged the local monopoly representative with fraud, specifically with failing to include all the region's tobacco farmers on the *tazmias*. As of early 1773, however, there was no determination that an actual fraud had occurred. Writing about the episode, another Cuban official noted that the *factor* had explained his actions and was reminded that he should be more careful in his duties so that all the tobacco harvested was indeed collected. Nevertheless, by mid-spring, don Martín de Asanza was named as his replacement.[34]

The idea that the island's colonial administrators were directly involved in contraband was not beyond reason. According to J. R. McNeill, "Virtually every prominent official in Cuba in the early eighteenth century was accused of complicity in illicit trade, and at least one spent years in prison in Santiago de Cuba for his activities." Even into the nineteenth century, the situation remained unchanged, as (in the words of Quiroz) "colonial corruption was a consequence of the systemic weakening, and at times, violent destruction of the local institutional bases of Cuban autonomy through discrimination, racial checks, war, expropriation, and exile." Cuban officials were among many royal employees in Spanish America engaging in illegal activities involving tobacco. For those in Costa Rica's tobacco monopoly, low pay for the tobacco police prompted them to join the *vegueros* and engage in contraband. Various tobacco monopoly bureaucrats in Mexico committed fraud and embezzlement against the *estanco* there, including one case involving eighteen thousand *pesos* in cash and five thousand *pesos'* worth of manufactured tobacco goods.[35]

Tobacco contraband or fraud was nearly impossible to resist for those with the disposition to engage in such types of illegal activities. Todd Butler observes that one of the "closest advisors" of England's King Charles I (reigned 1625–1649) was Sir Thomas Wentworth, the Earl of Strafford, who had charges of treason leveled against him for creating his own "corrupt system of tobacco licensing in Ireland." The formal indictment before Parliament, as described by Butler, alleged that "Strafford's system bore a significant resemblance to royal policy in England—trade was forbidden without a license, and Strafford imported 'divers great quantities of Tobacco' while simultaneously forbidding other merchants from trading with anyone but himself and then only at prices he declared." The charges, which could have been written in such a way that it implied that the earl acted on behalf of the monarchy, instead focused on Strafford's attempts to enrich himself and advance his authority.[36] Even for those in the highest positions of authority and connection, the illicit trade and activities surrounding tobacco proved irresistible.

Spain did not give up on its attempt to control tobacco on the island or limit its sale. A royal order of March 1787 reinstated the *resguardo montado*, and the group was formally reestablished on 12 November. One corporal would serve as its head, with five guards as subordinates, all to be stationed in Havana's barracks. The only member from the old unit who was not in prison, don Gonzalo Blanco Sacido, was made second-in-command. This time, the men for the new unit were scrutinized closely and chosen for their loyal service to the Crown. For example, guards don Manuel de Arza and don Manuel de Miguel y Pérez had participated in the disastrous Algiers expedition and the successful attack on Pensacola. Cuba's intendant proposed don José Miguel, the head of the *resguardo a pie mas antiguo* (the tobacco police who patrolled on foot), as the new head of the *resguardo montado*. Yet, not everyone agreed with this suggestion. In a letter signed by José de Gálvez, Raymundo de Onís, the head of the port and Miguel's former supervisor, objected to Miguel, arguing that he "does not have sharpness, nor the zeal that made him stand out in the Administration of the Port, leaving the post without notice from the intendant."[37] Don Ramón Bretos was appointed to the position. As the new century approached, however, his health problems became insurmountable and, despite seeking the services of a bloodletter, he was dead by early February 1800.[38] An examination of employment records indicates that the mounted tobacco police saw another serious reorganization in July and August 1805, although the cause for the introduction of five new members is unclear. In contrast to most of the founding members of the unit, these hires were apparently able to stay out of trouble, as they were all still employed by the monopoly at the end of 1808.[39]

A few Cuban authorities did distinguish themselves through their attempts to prevent illegal activities, particularly contraband. One such person during the 1770s and 1780s was intendant Juan Ignacio de Urriza. James A. Lewis describes Urriza as "an acerbic, honest and ambitious official who instinctively knew which laws to enforce in order to gain favor at the Spanish court."[40] Similarly adept at proving his worth as a loyal and accomplished bureaucrat, Urriza successfully petitioned the Marqués de la Torre to give the post of *capitán del partido* additional police powers.[41] As discussed in Chapter 5, Urriza informed Captain General Navarro about contraband tobacco seizures no fewer than seventeen times in 1779.

WHILE CONTRABAND PROVIDED A CLEAR, visible target for much of Spain's economic and political efforts during this period, it was not the only avenue undertaken by those wishing to circumvent royal authority. Another method at their disposal was failing to follow monopoly regulations properly. Careless packing of tobacco destined for the Real Factoría de Tabacos led to a damaged product, with its value either lessened or destroyed. In a letter dated February 1754, Cuba's captain general complained about the situation to the governor of Santiago de Cuba, Alonso Arcos y Moreno. Cagigal de la Vega indicated that much of the tobacco from that area was packed and shipped while wet. Tobacco in such condition was in violation of the official contract and could introduce humidity that might harm the entire shipment. The captain general openly accused those involved of "indolence" by knowingly dampening the leaves before selling them to the monopoly. Such actions were not only "beyond whimsy but went against natural reason," Cagigal de la Vega fumed, and clearly were part of an attempt "to avoid the exportation of admissible" tobacco. The capital general again reminded Arcos y Moreno less than five months later that the area's tobacco farmers were expected to turn in seasoned, clean, and separated tobacco. The monopoly continued to have problems with the *vegueros* from eastern Cuba as late as 1804; one observer noted that some farmers had attempted to defraud the *estanco* by wetting the leaves, which gave them more weight but also caused them to rot.[42]

Spain knew that problems existed with tobacco shipped not only to the metropole but also to other Iberian dominions, particularly New Granada. Frequent complaints about Cuba's poorly packed tobacco during the early 1770s indicated that the issue went far beyond *veguero* resistance or inattention. On 9 October 1771, Captain General Bucarely informed Minister of the Indies Julián de Arriaga that the tobacco sent to Cartagena had been deemed unacceptable. The viceroy in Santa Fé de Bogotá had described the tobacco delivered from Cuba as being of the most inferior quality, Bucarely communicated to the high-ranking Spanish official. The problem remained uncorrected as late as the summer of 1774. Cartagena's new tobacco monopoly administrator, don Santiago de Viana, criticized the poor quality of the tobacco he received from Cuba, calling it "bad tobacco, of waste class." Arriaga subsequently reminded Bucarely's replacement and Cuba's new captain general, the Marqués de la Torre, of his duties. The island should send the type of tobacco requested in order for the monopoly to develop as much as possible, Miguel de Múzquiz stated in a letter to de la Torre in June 1774.[43]

Royal officials, however, directed their harshest language against those in Cuba when Spain itself began to receive poor-quality tobacco. Minister of Finance Miguel de Múzquiz's tone was comparatively neutral when he informed Captain General de la Torre that eight hundred pouches of ground tobacco sent to Seville's factory on board the brigantine *el Príncipe* was of bad quality, which had been the case with previous shipments. The following month, however, Múzquiz was blunt: "The greater part of the ground tobacco that has been sold in the past year, 1771, and in the current one of 1772 from the Factoría of that island to the factories in Seville have been declared by the experts of the same factories, and those of the General Store in Madrid, as entirely useless." Múzquiz's complaints about receiving poor-quality tobacco continued into the following spring. For his part, de la Torre attempted to rectify the situation, reminding his subordinates that tobacco should not be accepted unless it fell under the specific guidelines required by the monopoly. The middle of July brought additional objections from Spain, however, this time directly from Seville. Tobacco shipped to the city from Cuba via the port of Cádiz was damaged, factory officials claimed. Monopoly representatives noted during the off-loading and transfer process in Cádiz that the ships coming from Cuba were not properly ventilated. Without appropriate air circulation, any moisture entering or already inside the hold could not escape, and as a result, the tobacco might be harmed. This observation notwithstanding, Cuba's tobacco and shipping issues remained unsolved. The situation was so severe that by late December 1773, the principal *estanco* administrator in Cádiz refused to allow ships from Cuba to unload their cargo.[44]

The issues experienced by Spain with Cuban tobacco appear on the surface to have been ones related to shipment difficulties, not intentional negligence or attempted fraud. Subsequent complaints, however, point to the possibility that other forces were at play. De la Torre wrote to Múzquiz on 8 October 1773 that he could not understand why the tobacco shipped from Havana to Seville had decreased in weight by 24,618 pounds while in transit. When de la Torre looked into the matter, a local official in Havana assured the captain general that he too was not responsible. No one stepped forward to claim responsibility for the situation, from the lowest levels of Cuban bureaucracy through to the captain general himself.

Madrid continued to lodge complaints with the island's top official a year later. Múzquiz's reports indicate that the shipments to Seville's Real Fábricas were worse than ever. Tobacco imports, Múzquiz noted, now contained "strange leaves . . . and others from tobacco ordered burned" as well as green or uncured tobacco leaves. After months of relatively restrained communication on the topic, Múzquiz was direct and to the point. "What motive is there for such type [of actions] against what is contracted, and orders communicated?" he fumed.[45] By the monopoly's own regulations, all tobacco was inspected at numerous levels, with monopoly representatives ordered to deliver only a high-quality product. The continued inferiority of the tobacco shipped and the large discrepancies between its listed weight in Cuba and its actual weight on arrival in Spain suggest that the Real Factoría employees on the island were deliberately defrauding the *estanco*, especially considering the time that had gone by since the finance minister's initial message.

Another form of sabotage was to reduce the number of plants that could be harvested in the first place. In 1805, for instance, a Cuban official noted that some *vegueros* had abandoned their plants, and others had cut back production.[46] Open, violent measures such as the destruction of the plants in the fields were less common. One oft-repeated story describes how a farmer in Güines, José de Coca, burned eleven *vegas* in 1792 to protest the Real Factoría's destruction of fifty-eight *cargas* (loads) of poor-quality tobacco, the crop for the entire year.[47] Tobacco farmers in other areas launched similar protests against governmental authority, often as part of a nascent revolutionary movement against the Crown. Virginia planters turned to crop destruction on several occasions, first in 1682 and later in 1732.[48] Mexican tobacco farmers in the early stages of the nineteenth-century Mexican independence movement did this likewise. Under rebel leader José María Morelos, tobacco fields in the Veracruz province were burned as "part of their military strategy to destroy the Crown's resources."[49]

SPAIN'S ABILITY TO CONTROL and regulate economic activities within its colonies reflected the Crown's strength and stability. Threats such as those posed by traffickers in contraband tobacco tested the very essence of state control and demonstrated effectively that individuals operating outside of official, legal channels could not only challenge the state, but also successfully defy it. The tighter restrictions pursued by the authorities only served to fuel Cuba's illicit commerce in tobacco. Contraband never gained legal respectability, however, despite its long-standing presence in commercial activities in the Atlantic world. Rather than being viewed as clever entrepreneurs who were interested in participating in a profitable economic activity, pirates, privateers, and smugglers continued to be seen as extralegal actors who deserved neither the spoils of their labor nor protection under the law.[50] Why, then, was this illegal activity still prevalent during the late eighteenth and early nineteenth centuries? Faced with the possibility of harsh sentences, fines, confiscation of their crop, and other penalties, what motivated the island's *vegueros* to operate outside of the officially sanctioned monopoly?

The farmers' exact rationales are difficult to establish, yet archival research points to economic reasons as possible factors. The *vegueros*' compliance with the law rested heavily on the Real Factoría's ability to provide them with a profitable, or at least adequate, alternative to contraband. Paying the farmers promptly and establishing competitive official prices for the tobacco harvest were two essential components of this strategy.

As initially discussed in Chapter 5, however, it was in these two areas, however, that the monopoly frequently failed. The Real Factoría required Cuba's *vegueros* to present their tobacco in a timely manner, yet it often did not pay them expeditiously, causing some farmers to sell their tobacco through unauthorized channels. This was a particularly acute problem in Sancti Spíritus. As early as August 1766, Captain General Bucarely was alerted to the occurrence of fraud and the sale of tobacco to individuals, each prohibited under the official monopoly; in his report, Felipe Bellos told Bucarely the monopoly's failure to pay for the tobacco it received was to blame. Two years later, the problem remained uncorrected, and Juan Gutierrez informed the captain general that farmers were selling tobacco to non-monopoly sources because of "lack of money." The monopoly remained behind on payments to Sancti Spíritus' tobacco farmers even into the next decade, and in July 1775, Manuel del Portillo complained to the Marqués de la Torre not only about the payments being in arrears, but about how this was not the case with other locations throughout the island. The local *vegueros* were unhappy about this situation. In a second letter to de la Torre that month, del Portillo remarked that the monopoly's board had located funds to pay the area's *vegueros*.[51]

Holguín's *vegueros* at the far eastern end of the island also complained about not getting paid when they delivered their crop to the monopoly. Correspondence between various regional officials and the captain general in Havana reveals that the problem may have been not with the *estanco*, but with its *factor*. Antonio Luque, the lieutenant governor of Holguín, accused the city's *factor* of delaying payments to the *vegueros*. Cuba's captain general, the Marqués de la Torre, was informed about the situation by the summer of 1774. De la Torre responded that Luque's only recourse was to make the *factor* a scapegoat in order to address the *vegueros*' complaints, in the hopes of preventing further criticism from them. De la Torre also issued a notice to the island's lieutenant governors, stating that they should refrain from "the economic and governing management of the Factorías." The case against the *factor* was still active as of September of that year.[52]

Tobacco farmers in Bayamo also suffered from lack of prompt payment for their crop. In the spring of 1772, the *vegueros* were simply given papers stating that there was no money to pay them for their crop.[53] A year later, the Marqués de la Torre noted that there were profits available in the Real Factoría and suggested that the area's tobacco growers be paid from these funds. Issues regarding payment of the *vegueros* continued, however. The following summer, de la Torre wrote to the lieutenant governor of the region: "I obligate you to solicit the permission to find in this area some quantity of money, liberating them against the Treasury of the Tobacco Monopoly of this city, so that the *vegueros* of that jurisdiction, who only with the product of their crops can alleviate the miseries that they suffer, are contented in some manner."[54] No progress had been made on the issue by that fall, however. Even into the 1800s, Bayamo's *vegueros* could not count on prompt payment for their crop. Funds destined for the purchase of the region's tobacco in 1793 were delayed when two of the dragoons transporting it fell ill, one in Puerto Príncipe and the other in Villa Clara. The area's governor general finally learned on 1 October 1793 that the funds had arrived.[55] In 1808, a *veguero* from Bayamo complained that monopoly officials often gave farmers promissory notes when they turned in their crop because the local monopoly's account was in arrears.[56]

The view of contraband as a viable option may have also stemmed from the *vegueros*' belief that the Crown did not adequately compensate them. While some farmers may have been driven by a desire to maximize profits on their crop, other *vegueros* were unable to support their families from tobacco production, and officials often cited this as a chief motivating factor for those who turned to the contraband trade. The financial need was so great in some areas that local authorities, especially in Pinar del Río in 1768 and 1780–1781, petitioned the captain general to give surplus meat to the *vegueros* to help sustain their families. Cuban officials argued for decades that if some of the financial incentives for turning to illegal trade had been removed through prompt and fair payment by the *estanco*, the order demanding that the *vegueros* deliver all their tobacco to the royal warehouses might have been easier to enforce.[57]

The argument that contraband was an attractive economic option for the poor is not novel.[58] Economic need alone, however, is insufficient to explain the persistence of Cuba's contraband tobacco trade, particularly since Cuba's *vegueros* were not all from a single socioeconomic group or background. Archival research reveals that despite the royal tobacco monopoly's apparent rigidity, at least on paper and as portrayed in traditional studies, Cuba's *vegueros* exercised significant personal agency in their economic activities. The illegal nature of some of these activities represented both a financial threat and a potential challenge to state control in eighteenth century Cuba. The prevalence of extralegal behavior, described in numerous documents, demonstrates the ineffectiveness of Spain's tobacco monopoly on the island. This economic policy was particularly difficult to enforce after the unprecedented openness enjoyed during the ten-month British occupation of Havana. The participation or at least tacit acceptance of officials and others charged with preventing illicit trade in tobacco ensured that the monopoly would fail in its ultimate goal of securing all the island's best tobacco for its own use. Despite these hurdles, Cuba's tobacco industry proved to be a significant source of income for the Crown, as well as a vital component of the island's rural infrastructure during the late eighteenth and early nineteenth centuries.

7 "Shipments of Havana tobacco not arriving due to the present war"

THE ATLANTIC WORLD IN THE late eighteenth and early nineteenth centuries was a place of change and possibility. Where there were once colonies, new nations emerged. Neighbors became trading partners, and economic competitors developed into new sources of needed goods. The year 1763 saw the Atlantic world emerging from the first great world war. European nations and their colonies—settlements as far flung as New France, India, Martinique, and Havana—struggled to retain control of territory at home and abroad.[1] The Peace of Paris ended the Seven Years' War, but continued turmoil was on the horizon.

In the ensuing six decades, political and economic changes would transform the Atlantic world. The thirteen North American British colonies would become the United States of America. France's crown jewel in the Caribbean, Saint Domingue, became Haiti, the second nation in the Western hemisphere to earn its independence and the first black republic at a time when much of the Atlantic world still embraced slavery. In the midst of all this change and chaos, tobacco consumption and trade continued virtually unabated. European wars frequently manifested themselves as Atlantic Ocean shipping interruptions, resulting in nontraditional trading partners stepping in to provide much-needed products and supplies.

The amount of tobacco entering Cuba's tobacco monopoly (and subsequently available for shipment to Spain) fluctuated wildly. Periods of increases were often followed by dramatic downswings. After Spain regained control of Havana from the British in 1763, tobacco production remained low, averaging 38,005 *arrobas* annually. Tobacco harvests by 1766 increased to over one hundred thousand *arrobas* a year, continuing on this trend for nearly a decade. Yet, officials were not pleased with the *vegueros'* commitment to the monopoly, lamenting that before the English invasion, the *vegueros* had turned in their tobacco to the Real Factoría without problem, but now they did not. The crop presented to the *estanco* rose even higher from 1772 to 1774, despite these concerns, with production averaging 255,088 *arrobas* each year. The annual yield from 1775 to 1778 dropped to the mid-100,000 *arroba* range, and dipped even lower in 1779 before rising again in 1780. Between 1781 and 1783, Cuba's tobacco crop entering the monopoly averaged approximately 40,000 *arrobas* annually. The high point of Cuban tobacco production under the monopoly from 1763 to 1803 was in 1788, when production peaked at 341,984 *arrobas* 11 *libras*. At no other point during this time did tobacco yields even exceed 297,000 *arrobas*. After the peak in 1788, each year generally saw less and less tobacco entering the monopoly's warehouses. By 1803, only 37,085 *arrobas* of tobacco were collected by Cuba's tobacco monopoly. Demand for the leaf remained high, however, despite the transatlantic shipping problems caused by war with England, which forced Spanish colonial officials during the last part of the 1790s and first decade of the 1800s to turn reluctantly northward to Virginia to supplement their existing supplies.[2]

Meteorological occurrences such as hurricanes, droughts, or storms had the power to affect the island's tobacco crop, and sometimes did. They do not appear to have changed the overall production rates during this period, however. The years 1768 to 1778 saw at least eight hurricanes and two periods of drought, yet, as described above, tobacco production during this period only increased. An increase in the price paid by the monopoly similarly did not have a lasting impact on production. Production did increase in each of the five western *partidos* near Havana during the same year that prices were raised on the highest-grade tobacco; however, this trend was not sustainable. The *partidos* of Guane, Güines, and Govea experienced several instances when production actually fell in the year following a price increase. Some of the growth in tobacco production nevertheless can be explained by the introduction of new tobacco-growing areas on the island and the expansion of existing ones. Having grown no tobacco whatsoever for the monopoly between 1762 and 1765, Matanzas and Puerto Príncipe both entered the tobacco market in 1766, with Matanzas as the third-highest producer of tobacco in the western part of the island. Tobacco production in Güines and Xiaraco likewise climbed after 1766, with each consistently producing in excess of twelve thousand *arrobas* annually between 1768 and 1778.[3]

Among the many aspects of the tobacco plant's cultivation, purchase, and production, perhaps the most unpredictable component involved shipping. When the tobacco passing through Havana's royal tobacco factory was ready for shipment, the wooden *La Machina* crane located on the *Muelle de la Factoría* loaded it onto various types of ships, including *fragatas* (frigates) and *paquebotes* (packet ships), for the long and often treacherous Atlantic voyage to Spain. An attempt was made to avoid the hurricane season for the most part, with increasing numbers of ships carrying Cuba's tobacco shipments setting sail as the fall progressed.[4]

Tobacco's safe arrival to Cádiz nevertheless was not assured. The dangers inherent in maritime life were just one layer of the threat to tobacco shipments. For example, Captain Estevan Cruz of the *goleta Nuestra Señora del Carmen* reported that the ship had struck a submerged stick and begun to take on water approximately two leagues from Bayamo, Cuba. Of the 224 *tercios* (bundles) of tobacco on the ship, all but 40 were lost. Another hazard to transatlantic shipping was bad weather. Ships waited to set sail until the fall, when the greatest danger of hurricanes had passed, but weather continued to plague the safe transport of tobacco from Cuba to Europe. In late October 1768, a storm—most likely a hurricane—caused the loss of the schooner *Juan Antonio Candelario* in the mouth of the Guanabo River. On board the vessel were 112 *tercios* of tobacco, 30 of which eventually were recovered. In the fall of 1771, tobacco bound for the Real Fábrica de Sevilla was similarly lost when the *San Antonio*, also known as the *María*, sank along the Punta de Guarico, just up the coast from Baracoa.[5]

Like these natural dangers to maritime transport, wars in Europe and the New World influenced shipping patterns during this period. The eighteenth century was fraught with political and economic turmoil. The year 1700 marked a significant shift in the ruling powers of Spain and Europe, as one of Europe's most powerful royal lines, the Hapsburgs (also known as the Habsburgs), reached its end. Founded as an international dynasty that united the Spanish and French royal houses, the Hapsburgs' first ruler was Carlos I (reigned as king of Spain, 1516–1556, and as Holy Roman Emperor Carlos V, 1519–1556). As a grandson of Spain's Ferdinand II and Isabel I, and of France's Maximilian and Mary of Burgundy, Carlos was descended from some of the most powerful monarchs in European history. Carlos's possessions covered a vast territory, from the Netherlands to Austria, and from Naples to Sicily, along with Castile and Aragon; upon his abdication, he divided them between his son, Felipe, and his brother, Ferdinand. This division of the line's hereditary lands spawned the Spanish and Austrian branches of the Hapsburg dynasty. The Spanish Hapsburg Empire continued to grow throughout the two centuries that followed, as new lands were added in both Europe and the New World.[6]

Earlier Hapsburg rulers had struggled with financial crises; however, Carlos II's problems in the latter part of the seventeenth century were more of a personal nature. The Spanish monarch was feeble and in ill health, and he lacked an heir. Carlos II, described by one scholar as "a near monster and imbecile, the victim of two centuries of Hapsburg interbreeding," had failed to produce an heir despite two marriages.[7] Scholars now cite the Hapsburg tendency toward intermarriage as a likely source for this difficulty. Nine of eleven Hapsburg marriages over a two-hundred-year period occurred between close relatives, including "two uncle-niece marriages, one double-first-cousin marriage and one first-cousin marriage." Like his grandfather, Felipe III, Carlos II was the product of an uncle-niece marriage. Recent scientific studies demonstrate that Carlos II possessed an inbreeding coefficient not unlike that for someone who was the product of an incestuous relationship such as father-daughter, or brother-sister. A high degree of genetic similarity therefore resulted in medical conditions that most likely caused the king's impotence or infertility.[8]

An heir, however, needed to be chosen despite these difficulties. Three possible candidates existed among the French, Austrian, and Bavarian royal houses. Of the three, the least likely appeared to be France, particularly since Spain and France most recently had been enemies until the War of the League of Augsburg ended in September 1697. Carlos II's initial choice, Electoral Prince Joseph Ferdinand, unexpectedly died in February 1699 at the age of six. Carlos was forced to write a new will, and on 3 October the French candidate, Duke Philip of Anjou, secretly was named his official successor. Carlos II's reaction indicated that the decision was not an easy one. The monarch "who previously had opened the tombs of his ancestors in the Escorial seeking some sign, now broke down in tears at the thought that he was betraying Habsburg family traditions." Less than a month later, the last Spanish Hapsburg monarch died at the age of thirty-nine, leaving his kingdom to a French duke.[9]

In nearby France, Louis XIV soon presented his grandson, the young Duke of Anjou, as "the King of Spain" to an assembly present at Versailles. Philip of Anjou left for Spain on 4 December 1700. He was accompanied by his grandfather for part of the journey, and crossed into Spanish territory on 28 January 1701. By 18 February, the seventeen-year-old king, now known as Felipe V of Spain (reigned 1700–1724), reached Madrid, where he was greeted by a throng of enthusiastic well-wishers. For England and others who feared a union between Spain and France, however, a French royal on the Spanish throne simply would not do. The resulting War of the Spanish Succession (1702–1713), although costly in the number of deaths, did little to change the royal houses of Europe, and instead cemented the French Bourbons as the rightful heirs to the Spanish throne.[10]

By the end of the eighteenth century, continued wars between Spain and its European neighbors had taken their toll on intercolonial trade between Spain and its largest Caribbean possession, Cuba. The frequent state of war between various European nations played out on the sea and caused additional disruptions to Atlantic shipping patterns. Tobacco shipments were not the only product affected by these issues, which similarly plagued any industry that relied on the transportation of their product through Atlantic waters. A vessel destined for Cádiz in 1779 did not arrive there. A Baltimore, Maryland, newspaper reported that a corsair had seized a "large American vessel from Boston, laden with naval stores, a large quantity of tobacco, and some rum." The American ship had attempted to avoid capture through "stout resistance," but it was taken, and brought ashore in Algiers. Other ships during this period suffered the same fate. An American privateer, the *Mary and Elizabeth*, piloted by Captain Weeks, sailed from Philadelphia for Cádiz, and was seized and sent into port at Lisbon. On board were 172 hogsheads (86,000 pounds) of tobacco and 10,000 wooden staves.[11] Wartime meant that any resident or ship was under probable threat from enemy forces, yet even during periods of peace, shipping remained a frequent target.

Pirates or privateers (the latter in essence pirates operating under the official sanction of a royal power) prowled the open sea in search of prizes for capture. Merchant ships, heavily laden and frequently unarmed, were easy targets. In early January 1761, the *London Chronicle* reported:

> About 30 sail of ships have been carried in Brest, Morlaix, and St. Maloes in the space of 15 or 20 days, by our [Dutch] privateers. Capt. Delilie, of Dunkirk, has taken three of the above; two of which had on board 1200 hogsheads [600,000 pounds] of tobacco, and the other 400 hogsheads of sugar. Our orders for the purchase of tobacco in England are stopped, on account of many vessels from Virginia and Maryland being brought into our ports by privateers.

Such illegal seizures of vessels laden with tobacco were common, and almost expected. Tobacco was a valuable commodity and easily found a market no matter what port in the world it entered. The actions of privateers, such as those described above, even altered normal purchasing patterns in areas like Flanders and created new sellers for this valuable product. No longer did the area's residents need to purchase their tobacco from London. Instead, it could be brought directly to them by their illustrious sea captains. The *London Chronicle* in mid-January 1761 again stated very matter-of-factly, "There have been taken, in the space of two months, fifteen ships from Virginia and Maryland, bound to the Port of London, having on board upwards of seven thousand hogsheads [3.5 million pounds] of tobacco; there are still missing from Virginia and Maryland, to the Port of London, seven more vessels."[12] Conflict made it difficult for traditional Atlantic trading partners to conduct business, but numerous opportunities arose for new trading alliances and suppliers.

The events leading up to the American Revolution (1775–1783) are well known, but Spain's contribution to the war and the impact of the war on transatlantic shipping during that time merit more study.[13] Calls for foreign assistance in the war between the North American colonies and Britain began as early as May 1776. British-born General Charles Lee, serving as representative for the Virginia Committee of Safety, wrote to the Spanish governor of Louisiana, Luis de Unzaga, requesting supplies from Spain. The provisions needed by the Anglo-Americans were weapons, blankets, and medical supplies, especially quinine, Lee indicated. By mid-summer, the Spanish delivered the requested goods to New Orleans. Spain and the North American colonies both feared continued British presence on the continent. Lee reminded the Spanish governor that if Great Britain succeeded in bringing the North American colonies back in line, the royal army and navy could easy turn their attentions to Mexico and Cuba.[14] It was an astute argument, particularly in light of the history between Spain and England, with the humiliating siege and ten-month occupation of the Cuban capital during the last years of the Seven Years' War still in Spanish memories. Lee's words certainly proved effective: Spain continued to provide aid to the Anglo-Americans despite its neutral status, and "by 1779, Spain had become an unofficial, de facto ally with the United States."[15] Early in the spring of 1779, Spain cemented its alliance with France with a treaty signed at Aranjuez. Spain refrained from openly siding with the North Americans at this point, yet one scholar has argued that this measure nevertheless meant "without becoming an ally of the rebelling colonies, Spain tied itself to independence."[16] Spain's intentions became clear on 21 June 1779, when King Carlos III announced that Spain would side with the colonies.[17]

The Atlantic Ocean during this time was the "busiest ocean highway in the world."[18] A heavy British naval presence in the Atlantic took a severe toll on shipping, only worsening what weather had already threatened, as noted later in this chapter.[19] Following Spain's declaration of war, Great Britain issued "letters of marque and a reprisal against the Spanish," a Boston newspaper recorded.[20] Less than two weeks after the start of formal hostilities between Spain and England, between eight and ten small vessels were seized off Bilbao in northern Spain. Tobacco shipments were so disrupted and had dropped to such a degree by the fall of 1780 that an official at Seville's tobacco factory was advised to eliminate some of the workforce, and to sell fifty-six mules and one horse. The animals, housed at the facility, were responsible for providing power to the factory's many grinding mills.[21]

The war raged on, as did demand for the plant, and Spain needed tobacco to continue to operate its tobacco factories on both sides of the Atlantic. Just as the thirteen colonies provided a source of flour for Cuba, they also possessed tobacco as well. Some in Spain worried that the lack of leaf would cause the monopoly to collapse. War prevented shipments of Cuban tobacco from reaching the peninsula, but the rebelling North American British colonies offered another option, particularly in the late 1770s and early 1780s. The patriot colonists were not the only ones who understood the value of the tobacco trade, however. Some in Britain realized the importance of commerce, particularly in tobacco, between the colonies and the metropole, even before the American Revolution's armed hostilities began. England's House of Commons speculated on the possible effect of a war on this trade on 5 January 1775:

> It is computed by those who are equally knowing in calculation as with the whole of the trade between England and America, that *every fifth man here* [italics in original] is some way or other employed in the commerce of the two countries, so as to derive his sustenance from it. What amazing loss then does this nation sustain by the positive expenses incurred by the soldiery and its consequences in America, as well as by the idleness of so many individuals at home![22]

Unable or unwilling to prevent the war that loomed on the horizon, Britain continued along the path that would lead to a rupture with its most important trading partner. The House of Commons again noted that the devastating loss of the colonies would be the gain of other nations on 9 May 1778. The English governing body bemoaned, "Tobacco is the staple of America, which the French, and it is believed the Spaniards, have agreed to take from them."[23] American residents were well aware of these sentiments in England's Parliament, as discussions from the House of Commons regularly were published in American newspapers.

Trade in tobacco between the North American colonies and Spain was not assured, however. Several obstacles existed. Virginia planters resolved to curtail their tobacco production as early as July 1774, with residents of Fairfax County resolving not to plant or cultivate tobacco "if American grievances be not redressed before the first day of November."[24] This presented the possibility of not only a major loss for tobacco consumers, but for the English Crown as well, with an American newspaper noting the House of Commons' belief that "tobacco is a very principal article of the revenue of Great-Britain."[25] Also, the Chesapeake region's tobacco suffered from perceived threats from enemy ships, along with its vulnerable location near the coast and accessible rivers. Shipping from the Chesapeake especially became difficult as British ships patrolled the area capturing any vessels that happened their way. Reports indicated that the British presence in the area was exceptionally heavy in the late summer and early fall of 1779.[26] An American newspaper informed its readers that by August 1780, at least nine vessels, five schooners, and four brigs, "mostly laden with tobacco, bound to France and the West Indies, have been lately taken in and near Chesapeake, and carried to New York," then held by the British.[27]

Despite these obstacles, no fewer than seven American-flagged ships appeared in 1780 and 1781 at the southern Spanish port of Cádiz. On board all but one of the ships were cargoes of Virginia leaf tobacco they hoped the monopoly would purchase.[28] Not just any quality of tobacco was accepted, however, as several episodes from the fall of 1780 demonstrate. Four separate ships arrived in the port carrying Virginia leaf tobacco during that time, but Spanish officials refused all four deliveries. The tobacco on board the *Paul Jones*, the *General Arnold*, the *Virginia*, and the *Necessity* was of "infamous quality and bad taste," Spanish officials asserted.[29] Although Spain clearly needed tobacco, not just any quality would do. Spanish officials noted that Virginia tobacco was inferior to that from Cuba, but in times of need, it was better than nothing.[30]

Spain's interest in purchasing Virginia tobacco to supplement or even replace its own supply of the leaf was not unprecedented. Spain had entered into contracts to purchase Virginia tobacco as early as the 1740s. Throughout the pre-Revolutionary War period and beyond, even after the abolishment of Spain's monopoly on tobacco, multiyear contracts ranging typically from three to five years for the purchase of Virginia tobacco were not uncommon. A contract between Thomas Macon and the Spanish government, presumed to be from the late 1760s, details some of the conditions typical of these contracts. Macon agreed to provide tobacco for cigars to Seville and Cádiz for a period of five years, from 1 January 1767 to the end of December 1771. After signing the contract, he had four months to begin to supply tobacco, which was to be of good quality, and "fresh, mature, healthy and without fault or bad odor." Macon was not responsible for customs or other fees, as these costs were calculated into the price he would receive. Macon, the contract continued, would be given a key to each storehouse where he was to place the tobacco; certain Spanish tobacco officials, including a superintendent, an administrator, and an accountant, would have copies of these keys. These terms would remain the same, the contract clearly stipulated, whether a state of peace or war was in effect.[31]

The thought of Spain purchasing tobacco from its competitor, Virginia, might seem implausible, but the commonwealth had long found unexpected markets for her products. As early as 1681, Virginia's governor, Lord Thomas Culpepper, had lobbied for ways to open up the Russian market for Virginia tobacco. When the Virginians learned of the favorable stance of Peter the Great (reigned 1689–1725) toward tobacco, the colonialists redoubled their efforts. Together, Virginia and Maryland merchants in 1697 drew up a petition to England's King William III (reigned 1688–1702) to encourage the Russian czar to allow the importation and sale of tobacco. Some of the arguments that the colonial merchants used in support of their product included tobacco's popularity with soldiers, its fatigue-fighting abilities (particularly in cold climates), and the financial benefits to the Russian treasury in the form of import duties. Discussions were soon underway between the two monarchs. A contract was signed between the two countries, with the first shipment of tobacco to equal 3,000 hogsheads (1.5 million pounds of tobacco), and the second year's shipment at 5,000 hogsheads (2.5 million pounds of tobacco). The contract gave the English the exclusive right to import tobacco into Russia, with only the Ukraine allowed to bring in tobacco for domestic use. As would be the case with Cuban tobacco shipments to Spain, war interrupted the delivery of tobacco from the New World to the Old. The War of Spanish Succession, as well as the war between Sweden and the combined forces of Denmark, Poland, and Russia known as the Great Northern War (1700–1721), ultimately contributed to the plan's failure, as shipments were unable to get through.[32]

Shipping tobacco to other areas benefited not only the buyer but the seller as well. The fledging United States sent twelve thousand hogsheads in 1779 alone to St. Eustatius, located at the northern end of the chain of Lesser Antilles islands, in an effort to procure much-needed naval supplies. Later Spanish records indicate that from there, Dutch ships typically took the tobacco "with ease" from that island and transported it elsewhere, including to Spain.[33] When that was not possible or feasible, Anglo-American ships were ready to make the possibly risky voyage to either Cuba or Spain in order to deliver their cargoes. Yet, Anglo-American ships were also subject to pursuit or capture by privateers, as noted earlier. These interactions were not haphazard accidents, but rather well-thought-out and well-planned attacks by the British or other powers. European authorities and even letters of marque supported their activities, and privateers plied the waters off the eastern seaboard of the former thirteen colonies, waiting to strike. These high-seas encounters and their chief participants were well known to those residing in the newly formed United States during the early 1780s. Day after day, readers of local newspapers up and down the Atlantic coast learned the fate of vessel after vessel as they fell prize to foreign ships.

American ships were prime targets for British vessels, such as the two US-flagged vessels seized off just off the coast in 1783. On 2 February, the *Independent Gazetteer* from Philadelphia reported that just weeks before, British privateers had separately pursued and taken the *Financier* and the *Nancy*. The *Amphion* and *Cyclops* had captured the Alexandria, Virginia-based *Nancy* on 17 January. The American ship, armed with six mounted guns (and designed to hold up to ten more), had contained a cargo of 120 hogsheads (60,000 pounds) of tobacco bound for Cádiz when it was seized. Similarly, the *Financier* had also carried a sizable load of tobacco—two hundred hogsheads (one hundred thousand pounds)—and it had left the Chesapeake in company of several other ships. Despite the presence of a convoy that included the thirty-six-gun *Emeraude*, a British man-of-war named the *Centurion* had claimed the American ship. The fear that must have come over the *Financier*'s crew must have been palpable as they sighted the vessel that captured them.[34] The British vessel was relatively new: it was a fifty-gun British man-of-war that had been commissioned on 10 May 1781. Under the command of Captain Samuel Whitworth Clayton, the vessel had participated in numerous captures along the former British colonies' eastern seaboard during the first year it was active.[35]

Threats from patrolling English ships along the coast were expected, albeit inconvenient. Yet, before tobacco could reach Spain's shores, after surviving a treacherous Atlantic crossing and avoiding enemy ships, it first had to leave American ports. The winter of 1779–1780 would present extraordinary hurdles for transatlantic shipping. Throughout the summer and fall of 1779, the weather provided hints of what the winter and the following year would bring. Rain, heat, and unexpected cooler temperatures all were recorded during this time. Crews traveling from the West Indies remarked that the islands were suffering from a severe lack of rain. No rain had fallen there from January to June of 1779, they reported.[36] While the Caribbean lingered in a drought, July began near Philadelphia with weather that was so unusually warm that it prompted Grace Growden Galloway, a wealthy loyalist, to declare by 6 July 1779, "Never was a Hotter spell of weather." The heat appeared even more dramatic when it was followed just several days later with what she described as two consecutive days of cold weather. In late July, the weather turned wet, with rain on 25, 26, and 28 July, and again on 2 and 6 August. The heat returned on Monday, 9 August, Galloway noted, describing the day as "violent hot." Yet, as of the middle of the month, there was little relief in sight Galloway confided to her diary on Saturday, 14 August, that "this is the Most hot oppressive day I ever felt." The following week contained much of the same. "Weather very Hot & wet & very Oppressive & Unhealthy," Galloway observed on Friday, 20 August.[37]

The fall of 1779 proved particularly windy for the middle colonies and New England. Newspapers noted a gusty September and October. A small privateer squadron intending to make port at Boston Bay was preventing from doing so by a "gale of wind" that "frustrated their Intention and now they are returned to refit," a New York City newspaper reported in mid-September.[38] Another paper, the *Royal Gazette*, commented that several ships had departed New York "in a very heavy gale of wind, which lasted until Friday the 17th, and drove them to the Capes of Virginia."[39] To the south, near Charleston, a bout of bad weather had made the transport of cannon and mortar from the American fleet difficult, leading to the failed patriot assault on Savannah.[40] The challenging conditions, however, did not prevent American ships from leaving ports and traveling across the Atlantic Ocean. Spanish officials in Cádiz reported in late October that two separate Anglo-American ships had arrived, each carrying Virginia tobacco. The schooner, *Venture*, was from Boston, while the brigantine *Patty* had sailed from Philadelphia.[41]

Residents of the East Coast were unprepared for the impending winter's unprecedented weather conditions, however. Cold temperatures, snow, and ice combined to plague the region from north to south. By Christmas 1779, it became clear that this winter would be particularly harsh. Severe weather forced a squadron of five ships under the command of French admiral the Count de Grasse to divert from the Chesapeake Bay and head out toward the West Indies.[42] Between 28 December 1779 and 7 January 1780, three separate snowstorms hit the mid-Atlantic and New England regions.[43] The leader of the Continental Army, General George Washington, was encamped near Morristown, New Jersey. Washington wrote in his diary that two earlier storms already had limited travel via local roads by the time a third storm began on 6 January. "The Snow which in general is Eighteen Inches deep, is much drifted—roads almost impassable," Washington noted.[44] Other nearby observers described the snow as significantly deeper, making it probable that Washington's measurement accounted for only the newly fallen snow.[45] Dr. James Thacher, a surgeon for the Continental Army, reported snowdrifts "from four to six feet deep" in early January.[46] These snowy conditions stretched throughout the region; in upstate New York's Genesee River Valley, snow was reportedly almost five feet deep.[47]

Yet, the severe conditions were not limited to the north. In a report on Williamsburg, Virginia, dated January 8, a writer exclaimed, "For a week past we have had the most severe weather that has been felt in this country upwards of forty years."[48] Between 1 January and 23 January, in Hanover, near Richmond, it snowed on three out of twenty-three days. Local temperature readings indicate the severity of the weather in Virginia's piedmont and tidewater areas. Temperatures in Hanover reached thirty-two degrees or higher at 9:30 a.m. on only three occasions during January. The lowest morning temperature reading was on 13 January, when the temperature dropped to just nine degrees, a dramatic decrease from a relatively balmy thirty-eight degrees just two days prior. Perhaps most revealing, these observations indicate that temperatures did not rise significantly during the day. During the first three weeks of January 1780, maximum-recorded temperatures in central Virginia reached into the forties on only four occasions.[49] A resident near Williamsburg, Virginia, noted similarly frigid temperatures between 7 January and 14 January. During that period, the high was a maximum of thirty-four degrees, recorded between noon and 3 p.m. on 11 January. Just a few days later, however, the temperature dropped dramatically to a mere six degrees at 7:30 a.m. on 13 January, rising to only seventeen degrees by 3 p.m.[50]

The extended periods of low temperature resulted in numerous frozen waterways. The *New-York Gazette* to the north reported on 17 January that the Delaware River "has been shut up by the Inclemency of the Weather for near six Weeks past," and in Philadelphia, ships were unable to navigate safely the icy waters. Spain's representative to the United States, Juan de Miralles, wrote from Philadelphia on 6 January 1780 that ice rendered all East Coast ports as far south as Baltimore impassable. The cold conditions only continued as the winter progressed. In early February, Miralles informed the minister of the Indies, José de Gálvez, that Virginia's ports likewise had frozen. Williamsburg, Virginia, reported "vast quantities of ice" on the area's rivers, which were preventing ships from reaching shore. Later accounts indicate that perhaps as much as twenty feet of ice accumulated along Virginia's coastline and remained until spring. At the end of January, a resident of New Haven, Connecticut, stated, "The eldest person among us, scarce ever remember, to have seen so much ice in the [Long Island] sound, as there is as present; at the westward it is entirely frozen across, and where it is wide, there is but little water to be seen."[51]

Maritime transportation came to a halt during this period, but enterprising New York City–area residents found new means of getting around. "Great numbers" of people, along with a sleigh and two horses, crossed the ice-covered East River, a local newspaper noted in mid-January.[52] Other residents walked over the frozen waterway from Staten Island to New York City. A 31 January 1780 newspaper report provides a vivid demonstration of just how thick the ice was. The paper noted, "An Ox was roasted on the ice on the River Delaware, opposite to Philadelphia, last Saturday Week."[53] Such occurrences became commonplace as the long, cold winter dragged into February. Along with a series of notes regarding recent war activity, the *New-York Gazette* on 7 February casually mentioned at the close of an article that "yesterday 86 loaded Sleighs went from this City [New York] to Staten Island on the Ice."[54] Ice crossings continued in the area well into mid-February. In the *Pennsylvania Packet, and General Advertiser*, a report on conditions in Connecticut said, "The intensely cold weather we have had for some time past, has frozen the western part of this sound for several miles—at White Stone, to such a degree, that people pass to and from Long-Island, on the ice."[55] Such feats were not limited to New Haven, Philadelphia, and New York City, however. Farther south, a Virginia infantry regiment was able to cross the Rappahannock River, frozen since the previous November, traveling from Falmouth south toward Fredericksburg on the far bank.[56]

The severe weather was not limited just to the former British North American colonies, as Cuba and the rest of the Caribbean also struggled with climatic events in 1780. To quote historian Sherry Johnson, "Historical climatologists rank the hurricane season of 1780 as the deadliest in terms of the loss of human life, estimating that over 20,000 people perished in the hurricane of 10–16 October in just one landfall in the Leeward Islands."[57] The season had started even earlier than usual, with a rare winter storm that had struck near Havana on 22 February and continued into the next day. A storm surge in excess of eighteen feet had threatened Cuba's fleet as naval commander Juan Bautista Bonet and his men "struggled to save the ships." Strong winds had prevented the ships from heading out to sea into safety, but in doing so had inadvertently saved the sailors' lives. Bonet later admitted that "in his fifty-three years of service he had never experienced such a strong gale in the winter. . . . If they had been at sea, all aboard would have perished."[58]

AS IT CONTENDED WITH WAR, patrolling enemy ships, and foul weather, Cuba found itself in desperate need of goods and supplies during the course of the

American Revolution. Disruptions to typical shipping patterns meant that the metropole "practically ceased to be a supplier and customer for Cuba."[59] Unable to procure supplies through Spanish-flagged ships, which were frequently the target of British privateers, the island turned to the United States. Anglo-American ships began to arrive in significant numbers in the last part of 1780. Much of this north-south trade focused on flour and other necessary provisions. This was not the first time, however, that Cuba had turned to the north for much-needed food supplies. The Creek Indians around the Tampa Bay area of Florida provided the island with *tasajo*, the dried jerk beef used as a main food source for Cuba's African slaves. The product was so in demand that the island's captain general had considered a proposal by the Creeks to have a ship outfitted for trade in *tasajo* and other goods (the proposal had been rejected, however).[60]

After the conclusion of the American Revolution, tobacco remained an important commodity sent between the New World and the Old World. Just one year after the formal end to the conflict, both American and English newspapers printed an announcement that speaks volumes of the desire to continue the trade in tobacco. In a declaration dated 16 April 1784, King George III declared from the Court of Saint James that "for a limited time, certain powers for the better carrying on of trade and commerce" would be permitted. As such, "any tobacco, being the growth or production of any of the territories of the said United States of America, may, (until further order) be imported directly from thence."[61] This opening of trade not only allowed Great Britain access to a much desired product, but it also provided a market for the United States' fledgling economy.

The former British colonies also continued their transatlantic trade with the Iberian Peninsula. A decade after the Treaty of Paris, in the late spring of 1793, a newspaper in Philadelphia noted that the *Enterprise*, piloted by Captain Jones, had arrived in the port the day before, back from Cádiz. Two additional ships—the brig *Minerva*, under Captain Thomson, and the brig *Mary*, commanded by Captain Wharton—had remained in Cádiz but were to leave within three and eight days respectively.[62] As of 1833, Spain was still entering into contracts for the purchase of tobacco from Havana, as well as from Virginia and Kentucky. All three types of tobacco were to be delivered under the conditions stipulated in the contract for a period of three years, beginning on 4 August 1833 and ending on 4 August 1836. The agreement called for approximately three million pounds of tobacco from Virginia and Kentucky to be delivered directly to the Real Hacienda, and specified that the tobacco should not be sent via Gibraltar or another European contractor.[63]

While there was a brief respite in hostilities after the end of the American Revolution, which allowed ships to cross the Atlantic relatively unthreatened, European powers were soon again at war. Thus, in the nineteenth century, capture by enemy ships remained a real possibility and disrupted transatlantic shipping. Cuban intendant Luis de Viguri noted in early 1800 that the Anglo-Spanish War (1796–1808) had made it difficult for ships leaving from Cuba to reach Spain. Viguri stated in his letter to Miguel Cayetano Soler, Spain's minister of finance, that he had considered sending a ship loaded with tobacco to the United States. There, it could take on "false papers as if it was American property" and then head to the first available port in Spain. While it is unknown what Soler's response was to this suggestion, the scheme was not without historical precedent. The island had resorted to foreign vessels for earlier shipments to the peninsula. A Swiss ship, the *Arditia*, had left Havana on 8 October 1771, carrying a load of tobacco destined for Seville's tobacco factory. Tobacco was not the only commodity affected, and any industry that relied on its product's transportation through Atlantic waters was likely to suffer losses as Spanish ships faced the very real possibility of capture.[64]

No matter its original source, or the flag on the carrier that had transported it, all the tobacco that safely made it to Spain followed the same path to its final processing in Seville. Tobacco typically entered through Cádiz, which in 1717 had officially replaced Seville as the main port of entry for all goods from the New World. Seville, located sixty miles up the winding Guadalquivir River, earlier had proved a natural port. Most oceangoing vessels between the sixteenth and eighteenth centuries were quite small, and had thus been able to traverse the river to that point, a route that protected them from attacks. As ships grew larger and heavier, however, and the Guadalquivir River silted up, the inland location became impractical. There was limited space at Seville's docks, which made the area dangerous for larger ships and their maneuvers. Yet, Cádiz possessed a few maritime challenges because of its own location. Sailors noted that the port suffered from strong southwest winds called *vendavales* during the winter months of November to March. Additionally, these winds were known to "bring heavy and constant showers, which conceal the land, and throw in so much sea that all the bars become impassable," along with a dense fog that lingered over the area.[65] Once tobacco reached Cádiz, Spanish tobacco monopoly officials inspected the product and decided either to accept or reject it. From the port city, tobacco traveled inland up the Guadalquivir River toward Seville, reaching the city in as little as three days, or as much as twenty-one days.[66]

The process remained much the same as time passed, although the increasing demand for tobacco products over the course of the eighteenth century rendered Seville's original seventeenth-century San Pedro *fábrica* obsolete, and plans to construct a bigger facility near the banks of the Guadalquivir River were underway as early as 1725. Construction on the complex began in 1758 and ended in 1770 at a reported cost of thirty-seven million *reales*. The building was massive in size, measuring 624 feet in length and 524 feet in width. It contained over two dozen courtyards and patios, twenty-one fountains, ten wells, a jail, and some churches, as well as houses for the superintendent of the tobacco monopoly and other key officials. When the tobacco from Cádiz arrived at the complex, it crossed a deep moat and then entered the main building through a columned façade. This façade featured depictions of ships, and of tobacco in all its states of production, and gracing the top of the edifice was the winged figure of Fame blowing a trumpet. Inside the building's first floor, over 110 horses worked to power the factory's numerous mills, which ground tobacco into snuff. Only men worked on this main floor, but upstairs, in a large hall that ran nearly the entire length of the building, it was mostly women worked rolling cigars. In all, approximately 1,400 people, including 1,300 day workers, oversaw the processing of the never-ending influx of tobacco from Cuba and, to a much lesser extent, the rest of the New World. In the form of two paintings by Francisco Goya, rendered at a cost of four thousand *reales*, His Royal Highness King Carlos IV and his queen, María Luisa de Parma, watched over the factory's daily activities.[67]

It is likely that the consumers of the product during this period spared little thought for the long and treacherous journey the leaf had made from its fields in either Cuba or Virginia to their local tobacconist. For some, however, the difficulty of getting tobacco from the grower to the user may have enhanced its status.

Epilogue

ONE HUNDRED YEARS AFTER Cuban tobacco was first placed under governmental control, the monopoly ended in 1817. King Fernando VII and his predecessors had focused their attention on "the promotion and cultivation of the precious plant of tobacco, so important for the exclusive privilege that nature has given the soil of the island of Cuba to produce the most exquisite quality in the world." The king now acknowledged that the "privileges given to the factory have been the cause of the decline of tobacco production, which had previously exceeded 600 million *arrobas*, and which today, according to the most impartial reports, does not reach half that amount." All the privileges enjoyed by the monopoly ended on 23 June 1817, and the cultivation and sale of all classes of tobacco was now completely free of government-imposed restrictions. News of the royal decree reached Cuba in mid-August, through a passenger on board a ship that had left Cádiz. The man carried a 3 July supplement to the *Gaceta de Madrid*, which included a republication of the decree. By the end of September, it had been reprinted in Cuba's *Diario del Goberino de la Habana* and the *Memorias de la Sociedad Economíca*.[1] Cuba's long governmental control of one of the island's best-known agricultural products had officially ended.

Spain's royal tobacco monopoly in Cuba had appeared to be an all-powerful colonial institution on the surface. It had survived a turbulent start marked by three uprisings, and remained in place for one hundred years. Tobacco and Spain's monopoly shaped Cuba, Spain, and the Atlantic world in dramatic ways during the late eighteenth and early nineteenth centuries. The bureaucratic structure of the monopoly altered the population characteristics of Cuba, and along with the expanding military presence on the island, led to a large Spanish peninsular population at time when most of Spain's colonies were becoming increasingly creole. In a society where class and race frequently divided the residents into near impassable categories, Cuba's tobacco industry drew nearly all the sectors of the island's population together.

Tazmias indicate that the existing historiographical image of the *veguero* as a poor, white Canary Islander is inaccurate at best. As this study has shown, people of all walks of life worked together in the island's tobacco industry from 1763 to 1817. Paying attention to the presence and importance of people of color in Cuba's tobacco production provides a better understanding of the labor force that contributed to bring this valuable crop to market. While the impact of tobacco-growing on the ability of slaves working in the fields to earn enough money to purchase their freedom still needs further exploration, it is clear that tobacco offered another avenue for Cuba's people of color. The plant may have even provided them with a way to integrate themselves further into the general population in terms of social and economic opportunities that would otherwise not have been available. While *tazmias* made note of slaves who labored in tobacco for their own benefit, they usually did not record the names, status, or other identifying information about each individual worker on the *vega*. Instead, they listed the number of workers under the general label of *operarios*. While one might reasonably assume that a *vega* listed as having one "worker including the owner" meant that the owner alone tended to the tobacco, it is difficult for us to know whether this was in fact the case, as an owner might have employed a hired hand or slave to labor in the fields.

The Cuban countryside during the late eighteenth and early nineteenth centuries was a diverse setting with many agricultural pursuits in which its residents could engage. The cultivation of tobacco was one of these, but as this study has argued, its effects went far beyond the symbolic association of Cuba and a good cigar. The *vegueros* who grew tobacco during this time represented the width and breadth of Cuban rural society—whites, blacks, freed people, slaves, lowly peasants, and those with noble titles. Their resistance to the royal tobacco monopoly—manifested in outright contraband, sloppy production habits, and demands for improved terms—demonstrates that the Bourbon monarchs were not all-powerful lords presiding uncontested over the island's economic life.

United in the cultivation of this popular crop, those working in Cuba's tobacco fields or sponsoring those who did were in agreement on their opposition to the monopoly. The government's tightening of control over so many varied sectors of the island's colonial populace through the implementation of the *estanco* led to a society that chose to resist Spanish domination primarily through passive means. Other forms of opposition to government control of the tobacco industry included disobedience through the growing and selling of tobacco on the contraband market, or more subtle means of resistance that could be explained away by the *vegueros* as accidental, such as failing to use care in packing the tobacco for shipment. Spain struggled to overcome many of these challenges, but contraband in tobacco in particular proved an insurmountable obstacle in various locations throughout Spain's overseas colonies.

The restrictive nature of the monopoly itself was ultimately its own undoing, and was at the heart of many of the farmers' complaints. Although the existence of the monopoly guaranteed the tobacco farmers a buyer for their crop, provided it was of good quality, the farmers had little job or economic security, and they were frequently subject to external forces beyond their control, such as disruptions to transatlantic shipping. The large profits garnered by the monopoly even after all its expenses were accounted for indicate that it could have easily paid the *vegueros* higher prices for their crop, while still keeping tobacco a profitable enterprise for the Crown. The monopoly's actions demonstrate that governmental officials viewed the *vegueros* simply as a labor force, rather than suppliers of a specialized product. It was the monopoly—not the experts, the *vegueros* themselves—that determined the number of plants that could be grown, the areas in which tobacco cultivation could be expanded, and on occasion even determined who could or could not grow tobacco. When these laborers felt that their efforts were not rewarded, particularly in terms of monetary compensation, they either sold their product to an appreciative buyer, as contrabanders, or simply turned in tobacco with little regard to its packing or other quality controls.

While the Cuban tobacco monopoly failed to secure its position as the exclusive buyer for the island's tobacco, it was successful in exemplifying the incredible capacity of the Bourbon Crown to generate revenue and create a new colonial bureaucracy. Many individuals in Cuba, Spain, and elsewhere owed their livelihoods to tobacco. These people included the mule pack drivers, who carried tobacco to the royal storehouses; the bureaucrats responsible for administering the monopoly; the members of the *resguardo*, who were responsible for enforcing monopoly regulations; and the workers in both Cuba's and Spain's factories, who turned the raw product into one ready for consumers. A sampling of *tazmias* from just thirteen *partidos* in one year, 1770, reveals that 1,243 people, both free and slave, worked in the tobacco fields of those districts. During the highest year of tobacco production under the monopoly, 1788, ten thousand heads of households across Cuba were dedicated to the cultivation of tobacco. In 1817, the final year of the monopoly, 8,775 people worked in the island's tobacco fields.[2]

Some argue that in its early days, tobacco had not been intended as a capitalistic, commercial venture for Spain, whereas it was for places like Virginia. This work has demonstrated, however, that Madrid took definitive measures to ensure that it partook of the potential revenue deriving from this plant. These actions extended far beyond the implementation of the monopoly. Tobacco was always purchased by the Crown, even during wartime (when, for example, transatlantic shipments of sugar were prohibited). Beginning in 1798, Spain also sought to protect tobacco growers from large landowners with a series of decrees. Cuba's captain general, Juan Procopio Bassecourt, the Conde de Santa Clara, was "authorized to use every means in his power to enforce the decrees, including the use of regular troops."[3] In doing so, he warned that "in the name of His Majesty he would take the most severe measures against any of them [the large landowners] who contravened his royal intent."[4] Preferential treatment was extended to the *vegueros* on other occasions as well. In 1801, an existing exemption on the payment of *diezmos* (tithes) for those who grew tobacco was extended by the king for an additional ten years. The captain general was named "the official guardian of their interests" to further protect the *vegueros* in 1803.[5]

One indication of tobacco's impact on Cuba's colonial economy was the sheer size of tobacco production. Between 1761 and 1803, 5,876,844 *arrobas* (almost 149 million pounds) of tobacco were produced throughout Cuba. Production peaked in 1788, reaching almost 342,000 *arrobas*, or about 8.6 million pounds. Production in Güines alone during this forty-three-year period totaled 706,440 *arrobas 8 libras*, or over 17.9 million pounds.[6] The estimated value of Cuban tobacco sent to Spain between 1760 and 1810 was one hundred million *pesos*, accounting for 25 percent of Spain's total revenue during that period. Since these figures do not take into account clandestinely grown or sold tobacco that did not reach the Real Factoría, this amount should be considered a mere starting point.[7]

The economic size and importance of tobacco to the Crown can also be seen in the personnel outlays and capital infrastructure that helped the monopoly to function. The vast amount of money invested in both salaries and physical capital indicates that the Spanish government considered tobacco to be a profitable enterprise. Salaries within the monopoly ran the gamut between several hundred *pesos* per year to several thousands. Even within the *resguardo*, salaries varied widely. Those in the *resguardo a caballo* earned more than those in the regular *resguardo* but were also responsible for maintaining the horse, rifle, pair of pistols, and sword assigned to them. The *cabo principal* (head corporal) of the *resguardo a caballo* in Havana and the surrounding area earned 730 *pesos*, and a *guardo a caballo* made 547 *pesos* per year. Members of the non-mounted *resguardo* earned significantly less. A corporal or a *guardo* in the non-mounted *resguardo* earned 360 or 240 pesos per year respectively. In all, salaries paid to those associated with the *resguardo* and in charge of protecting the monopoly's interests totaled 15,677 *pesos* annually, according to a 1778 report. The *director general de la renta* (general director of the monopoly), by contrast, had an annual salary of three thousand *pesos*.[8]

Another obvious expense to the Crown was the purchase of tobacco from the *vegueros*. Although farmers frequently complained about the prices they were paid for their crop—charges that dated back to the earliest days of the monopoly—the Spanish government's willingness to outlay sizable amounts of money demonstrated the value that it placed on tobacco. In 1723, two hundred thousand *pesos* of the money sent annually from New Spain to Cuba to cover the island's expenses (known as the *situado*) was designated for the purchase of tobacco, but by 1744, this figure rose to five hundred thousand *pesos*.[9] Clearly Madrid believed that it would not only recoup these expenses through the sale of tobacco, but make a respectable profit as well.

The financial expenditures associated with Cuba's tobacco monopoly were merely an economic drop in the bucket compared to the value of the capital goods owned by the monopoly. An end-of-year accounting of all Real Factoría-owned possessions throughout the island noted their value at almost 1,772,000 *pesos*. Its possessions in Havana alone accounted for over 1.5 million *pesos*. The inventory included the building housing the tobacco factory, assessed at 1,238,890 *pesos*, and 62 slaves valued at 18,000 *pesos*. The total value of properties and goods related to the tobacco industry was even higher, since in "many provinces there are no *casas del Rey* [king's warehouses]," and the necessary facilities were rented from private individuals.[10]

Even into the late nineteenth century, the presence of people of color in western Cuba's tobacco industry remained strong. According to data from 1899, people of color accounted for 17 percent of all landholders in Pinar del Río, and held 11 percent of the area's land. The vast majority, 75 percent, of properties rented by people of color in Pinar del Río were tobacco farms, and additionally most free black renters "seem to have come from that [tobacco] background," the 1899 census noted. Throughout Cuba as a whole, the tobacco industry had expanded, with both the number of farms and the number of workers tending them increasing after the monopoly ended. In the first ten years of free trade, the number of *vegas* rose by about 40 percent, to 5,534. One-quarter of these were in Cuba's central department, while 29 percent were located in the eastern department. As the nineteenth century continued, tobacco cultivation expanded throughout Cuba, particularly in the central and eastern parts of the island. Cuba held 9,482 tobacco farms by mid-century, with nearly 60 percent located in the eastern part of the island, from Las Tunas to Baracoa. Tobacco's impact was felt throughout the length of the island, not just in the areas typically associated with the plant.[11]

A series of natural disasters in the form of hurricanes, heavy rains, and droughts during the 1840s was particularly devastating to the island's tobacco industry. Even before the San Francisco de Borja hurricane of October 1846 hit, a resident of Guane, in the far western part of the island, complained that the damage to the area's tobacco industry was irreparable. The previous year's tobacco held in storage sheds and warehouses was destroyed, along with seeds for the next year's crop and young tobacco plants in the seedbeds. Out of seventy-six thousand plants on one *vega* in Govea, none survived, while seedbeds around Artemisa, Alquízar, and San Antonio de los Baños were wiped out. This was particularly devastating, as it followed on the heels of a hurricane just two years earlier, yet the island's tobacco industry "eventually emerged as another long-term beneficiary of the devastation wrought by the hurricanes of the 1840s." In western Cuba, the area's coffee plantations had suffered as early as twenty years before due to fluctuation prices for their product. Louis A. Pérez Jr. argues that as a result, tobacco and sugar became solidified as the dominant agricultural industries in the island.[12]

Tobacco dominated the rural landscape of Pinar del Río as the nineteenth century progressed, further cementing the region's status as a prime producer of the island's tobacco. During the inter-census period between 1846 and 1862, the number of tobacco farms in the western part of Cuba increased from 3,990 in 1846 to 6,381 in 1862, a rise of 60 percent. Destruction wrought by the War for Independence (1895–1898) left eight hundred *vegas* throughout Cuba destroyed, yet, as late as 1899, 72.2 percent of the tobacco grown in the entire island originated in Pinar del Río's 14,928 *vegas*. Whereas just 9.6 percent of all Cuban land under cultivation was planted in tobacco, slightly more than 42 percent of Pinar del Río's area under cultivation was in tobacco during this time. The period from the 1817 opening of the tobacco industry's opening to free trade until the close of the nineteenth century marked an explosion in the number of *vegas* found in the area, increasing from 632 to nearly 15,000, an astounding 2,262 percent growth in the region's farms dedicated to tobacco cultivation.[13]

At the close of the nineteenth century, Spain no longer held possession of its "Ever-Faithful Isle." A series of unsuccessful attempts at independence flickered early in the century, only to be extinguished before they ever caught on. Mid-century, however, saw the first real attempt at independence, with the Ten Years' War (1868–1878). The revolt remained limited to an isolated portion of the island, concentrated in Oriente. While most of the insurgents wanted to replace Spanish rule, there was little consensus among them as to what to replace it with, or who should decide Cuba's eventual fate. The premature deaths of many rebel leaders, the lack of material support from the Cuban exile community, concerns over the role that free and enslaved people of color should possess in a free Cuba, and a desire by some insurgents to accept annexation to the United States all contributed to the failure of the revolt. Another small uprising, the *Guerra Chiquita* (Little War; 1879–1880) followed close on the heels of the Ten Years' War, but Spanish forces soon put it down.[14]

Ambitions for independence continued to fester on the island, and by 1895, Cuba again erupted. This time things were different. The island's slaves won their freedom in 1886, and now multiple social, racial, and economic elements throughout Cuba supported the rebel movement. The Cuban population living in the United States, from Key West to Tampa and New York City, provided material and financial support for the movement, raising funds for the cause under the leadership of poet and patriot José Martí. By early 1898, the Spanish military was on the defensive and the Cuban rebels controlled the countryside. Louis A. Pérez Jr. makes a convincing argument that the Cubans believed that victory was imminent. Whether this assessment was true or merely an optimistic wish is of little consequence. The insurgents thought it to be the case, and this conviction influenced their reaction to American intervention after the explosion of the USS *Maine* in 1898, leading many on the island to argue that the Americans had snatched victory from them.[15]

The last war for Cuban independence (1895–1898) and the subsequent Spanish-American War (1898) devastated Cuba and its countryside. Incomplete reports indicate that the policy of removing Cuban civilians from the countryside and placing them in concentration camps led to in excess of 102,000 deaths.[16] The island's agricultural infrastructure was similarly devastated, with "over 100,000 small farms, 3,000 livestock ranches, 800 tobacco *vegas*, and 700 coffee *fincas* [farms]" destroyed by the conflict.[17] Despite the damage, an 1899 report produced by the US government on the commercial situation in Cuba clearly spelled out the continued importance of the island's tobacco industry: "Nobody can deny that the tobacco industry of Cuba is of the greatest importance for Cuba. The country has developed the industry to such a degree that it stands or falls with tobacco, so to say."[18]

Throughout the eighteenth century, the Bourbon monarchs attempted to transform interactions between their overseas territories and the Crown in order to consolidate and reinvigorate Spanish rule. Four concerns—commerce and the economy, the military, culture and religion, and administration—constituted the primary foci of the Bourbon reforms. These reforms were often interconnected, as their efforts to control trade and eliminate illicit commerce demonstrated.[19] As described throughout this work, the royal monopoly on tobacco dictated that Cuban tobacco farmers deliver their entire product to designated officials. While clearly this was an economic reform, the Spanish military system was also involved, particularly because illicit commerce continued and the crown was forced to address the issue of illegal trade, as this work has demonstrated. The use of *guardacostas*, charged with preventing the intrusion of foreigners intent on smuggling, and the formation of the enforcement bureau of the tobacco monopoly, the *resguardo*, were just two examples of the merging of economic and military elements during the Bourbon reforms.

Just as tobacco consumption reached all levels of world society as its use spread globally, the control of Cuba's tobacco monopoly infiltrated all levels of

Spanish governance. Immediate control of Cuba's tobacco monopoly rested with those on the island, but high-ranking government officials in Spain were regularly apprised of even the most minute details regarding the monopoly's decisions and day-to-day operations. Until his death in January 1776, the "aging, conservative, but venerable" Julián de Arriaga oversaw all monopoly activity.[20] Arriaga's position as head of the Ministry of the Indies since 1754 meant that he "theoretically exercised complete administrative authority over the colonies."[21] Arriaga's involvement as it related to the tobacco monopoly, however, was far more than just theoretical and included knowledge of virtually every aspect of the monopoly's operations. In just a two-year period, 1769 and 1770, this hands-on control included knowledge of contracts for Cuban tobacco mills, shipments of tobacco to Peru and the Canary Islands, meetings of Havana's tobacco monopoly board, and money sent from New Spain to the Real Factoría in Havana. Other details that did not escape Arriaga's attention included a fire in the Río de Sasa tobacco warehouse, located in the *partido* of Sancti Spíritus, and the negative effect lack of rain would have on the 1769 tobacco harvest.[22]

Other high-ranking members of Spain's court also were intimately involved with Cuba's tobacco industry and monopoly. The peninsula's ministers of finance frequently addressed matters related to tobacco. In late December 1775, Miguel de Múzquiz, the man who had replaced the controversial Marqués de Esquilache after his exile in 1766, reminded Cuba's captain general that "the matter of chief importance that the King can charge you with is that of maintaining the Factoría de Tabacos in that city [Havana], because that allows the Real Hacienda the ability to have the tobacco that it needs for the provision of this kingdom and those in the Americas." Múzquiz stated that the Marqués de la Torre needed to focus on this "while achieving at the same time" prices that would allow tobacco workers "the benefit of providing for their families." On 28 June 1800, a later minister of finance, Miguel Cayetano Soler, was informed that a Cuban *veguero* named Manuel Díaz who lived near Güines had petitioned the tobacco monopoly's board for an advance. Approval was granted to give Díaz the money he requested for the purchase of two slaves to help him with his tobacco crop.[23]

Yet, no one was more conscious of the value of Cuban tobacco as an important commodity to the Crown the king himself. In the summer of 1769, when Captain General Bucarely petitioned for additional slaves to work for the monopoly, this request was forwarded to Carlos III. Minister of the Indies Arriaga assured Bucarely on 22 September of that same year that he had spoken to the king regarding this request, and that the king had given consent for the purchase of twenty slaves. When don Manuel Ramos de Villavicencio, the man chosen by the *vegueros* in the Partido de la Canoa to serve as their *diputado* (elected representative), informed the tobacco monopoly board in early January 1770 that he was unable to fulfill his duties, Carlos III was also apprised. The construction of new warehouses, such as the one in the Partido de Mayarí, and the naming of a new warehouse guard along Cuba's southern coast (mentioned in Chapter 5) were all issues that crossed the Spanish monarch's desk.[24]

The preference for Cuban tobacco over product from other areas was not new. Cuba and tobacco have been inextricably linked throughout history. For tobacco devotees, only a Cuban cigar or other tobacco product will suffice. *Cigar Aficionado*, a US-based special interest magazine devoted to cigars and cigar smokers, launched in the fall of 1992 with the lead cover headline of "Cuba's Best Tasting Cigar." Between 2004 and 2011, thirty-five cigar varieties, all one-hundred-percent Cuban in content and manufacture, made the magazine's annual list of the year's top twenty-five cigars (see Table 8.1). The selections included cigars that were ranked among the top three brands on at least two occasions each. This distinction is even more impressive considering that it is illegal, and thus difficult (if not impossible) to import Cuban-made cigars into the United States. Additionally, Cuba has never had fewer than three cigars on *Cigar Aficionado*'s list in any single year. Cuba's record peaked in 2005 and again in 2006, with an impressive six cigars among the top twenty-five.[25]

Tobacco's popularity and the Cuban cigar in particular reflect the continuing Western interest in the unfamiliar. Tobacco marketing has used images of the exotic "other" throughout history. Examples range from the tobacco store's wooden Indian to the kilted Scottish highlander of eighteenth-century English tobacco shops.[26] Most Americans today can neither travel to Cuba nor legally purchase Cuban cigars, making the product even more exclusive, elite, forbidden, and craved. Therefore, enjoying a good Cuban cigar in many ways equates to a sort of virtual tourism.

Perhaps the most illustrative account of Cuba's close connection with tobacco and a quality cigar comes from Pierre Salinger, the former White House press secretary under US president John F. Kennedy. Salinger was a longtime smoker, but discovered Cuban cigars only after he started working with the president. Kennedy's favorite cigar was a Petit Upmann, and Salinger later stated, "Working around him, I felt I had no choice but to upgrade my smoke of choice to a Cuban." This shared love of a good Cuban cigar would prove helpful when Kennedy asked Salinger to do him a personal favor: deliver one thousand Petit Upmann cigars by the next morning. "Call all your friends who have cigars and just get me as many as you can," Kennedy said. Around 8 a.m. the following day, a successful Salinger entered the Oval Office not only with the requested one thousand prized Cuban cigars, but with an additional two hundred. The president opened a drawer of the massive, wooden HMS *Resolute* desk he was at and took out a long sheet of paper. "'Good!' he said," Salinger recalled in his 2001 memoir. He then quoted Kennedy

TABLE 8.1: Cuban cigars on *Cigar Aficionado*'s Top 25 List, 2004 to 2011

YEAR	CIGAR	RANKING
2004	Cohiba Siglo VI	3
	Montecristo No. 2	8
	San Cristobal de la Habana El Morro	12
	H. Upmann Magnum 46	15
2005	Montecristo Edmundo	2
	Partagas Serie D No. 4 Reserva	5
	H. Upmann Magnum 50 Edición Limitada 2005	14
	Cohiba Siglo II	15
	Montecristo D Edición Limitada 2005	17
	Bolivar Petit Corona	19
2006	Bolivar Royal Corona	1
	Romeo y Julieta Short Churchill	4
	Montecristo Robusto Edición Limitada	18
	Punch Super Selection No. 1	21
	Diplomaticos No. 2	23
	Montecristo Petit Tubo	25
2007	Montecristo No. 2	3
	Cohiba Maduro 5 Secretos	6
	Montecristo Petit Edmundo	11
	Punch Double Corona	16
2008	Punch Double Corona	6
	H. Upmann Sir Winston	7
	Juan Lopez Selección No. 2	11
	San Cristobal de La Habana La Fuerza	19
	Bolivar Belicoso Fino	22
2009	Cohiba Siglo V Tubo	2
	Partagas Serie P No. 2 Tubo	10
	Partagas Salomon	17
2010	Cohiba Behike BHK 52	1
	H. Upmann No. 2	9
	Partagas Lusitania	21
2011	Partagas Serie P No. 2	4
	Punch Double Corona	8
	Hoyo de Monterrey Double	15
	H. Upmann No. 2	21

Source: *Cigar Aficionado*, *www.cigaraficionado.com/top25*, accessed 4 April 2012.

as saying, "Now that I have enough cigars to last for a while, I can sign this!" Kennedy placed the document on top of the desk, and finalized an executive order outlawing all trade with Cuba, including the importation of cigars.[27]

Tobacco was frequently a source of internal conflict throughout history for those in high governmental positions. Kings, rulers, and presidents throughout the globe were all afflicted by the desire to extract its profit or regulate its use. For John F. Kennedy, Cuban tobacco continued to prove a tempting dilemma of conscience. Salinger met with Nikita Khrushchev in the Soviet Union in May 1962. At the end of their meeting, the Soviet leader told him, "Salinger, I see you like cigars. Well, I don't, and yesterday I received a wonderful present from Fidel Castro. I'm going to give it to you," Salinger later recounted. An assistant returned a few moments later with 250 cigars. Salinger worried that he would have difficulty getting them through US Customs, as they were illegal, but remembered he was traveling on a special, presidential-issued diplomatic passport. Both Salinger and the Cuban cigars returned without incident to the United States, and Salinger anxiously waited to share them with Kennedy.

The president, however, was not impressed. "Do you have any idea what a scandal it will create if someone finds out you brought those cigars illegally into this country, particularly since I banned them?"

Salinger replied, "It's no problem, Mr. President. There are only three persons who know about these cigars—you, me, and Khrushchev."

Kennedy would not budge. He ordered Salinger to take the cigars immediately to the head of customs, and demanded that Salinger bring back a receipt. "I don't trust you with regards to cigars," Kennedy informed Salinger.

A very dejected Salinger complied, and after accepting the receipt, inquired what would happen to the cigars. The customs chief stated that he planned to destroy them.

"Yes, I know," Salinger muttered, as he headed out of the office. "You're going to destroy them one by one."[28]

Glossary

alcalde	a local magistrate
almacen	a warehouse used to store tobacco
arriero	a mule pack driver
arroba	a measurement of weight, approximately equivalent to twenty-five pounds
burro	a piece of leather on which dried tobacco leaves were carried to the tobacco warehouse
caballería	a measurement of land, approximately equivalent to thirty-three acres
cabildo	a town council
capitán de partido	a district captain or constable
conuco	a plot of land used by slaves to grow crops for their own consumption or profit
contaduría	an accounting division
corral	a small estate
cuje	a stick of tobacco. Tobacco was placed on sticks and hung for drying and curing. The amount of tobacco placed on these sticks was also called a *cuje*.
estancia	a cattle farm
estanco	a monopoly; used in this work to refer to the tobacco monopoly

factor	a local tobacco monopoly representative or agent
factoría	a tobacco factory or monopoly
gavilla	literally "sheaf." A way of bundling tobacco in Cuba. Between twenty-five and forty-five leaves of tobacco were lined up with the central veins together and tied at the top. Four *gavillas* tied together were called a *manojo*, or bundle.
guardacosta	a sea patrol used to prevent contraband or other illegal activity
hacienda	an estate
hato	a cattle ranch
hijos	literally "children." The offshoots that developed from the base of the tobacco plant and were removed by farmers. They were usually known as "suckers" among US tobacco farmers.
ingenio	a sugar estate
isleño	Canary Islander
junta	a tobacco monopoly board
La Machina	literally "The Machine." A powerful crane located along Havana's docks, it was used to load tobacco onto ships.
manojo	literally "bunch" or "handful." In Cuba, individual tobacco leaves were stacked in piles (see *gavilla* above) and combined into larger bundles called *manojos*.
media vara	a measurement of length, approximately equivalent to half a yard
molino	a mill, where tobacco was roughly ground
Muelle de la Factoría	Monopoly Wharf. Tobacco was shipped off the island from this location.
operario	worker
Oriente	literally "East." This term refers to the eastern part of Cuba. In the early nineteenth century, Cuba was divided into three administrative districts: *Occidente* (Western), *Central* (Central), and *Oriente*.
padrón	census
partido	district. Within the three large administrative regions (see *Oriente*), Cuba was divided into smaller districts, each headed by a local *capitán del partido*.
rancho	ranch
Real Fábrica de Tabacos	Royal Tobacco Factory. This was a name for the tobacco monopoly.

Real Factoría	Royal Factory. This was a name for the tobacco monopoly and its associated buildings, particularly in reference to the location where tobacco was processed.
Real Hacienda	Royal Treasury
renta de tabacos	tobacco monopoly
resguardo	tobacco monopoly police. The names for the unit on horseback included *resguardo a caballo*, *resguardo de a cavallo*, and *resguardo montado*. The foot patrol was known as the *resguardo a pie mas antiguo*.
situado	a transfer of royal funds
tazmia	a period accounting document in which Spanish officials recorded the names and amount of tobacco grown by each farmer
temporal	a strong storm
tercio	a tobacco bale
vara	a measurement of length, approximately equivalent to one yard
vega	riverside plots of land where tobacco farms frequently were located in Cuba, and the usual term for a tobacco farm during much of the eighteenth century. The terms *vega* and *estancia* frequently were used interchangeably.
veguero	tobacco farmer

Notes

LIST OF ABBREVIATIONS

AFTS	Archivo de la Fábrica de Tabacos de Sevilla, Seville
AGI	Archivo General de Indias, Seville
AGS	Archivo General de Simancas, Simancas, Spain
ANC	Archivo Nacional de Cuba, Havana
CG	Correspondencia de los Capitanes Generales, ANC
Contratación	Casa de Contratación, AGI
DSC	*Documentos sobre Cuba* 1740–1806, vol. 27, LMC
exp.	*expediente*
GM	Guerra Moderna, AGS
IGH	Intendente General de Hacienda, ANC
LMC	Leví Marrero Collection, Special Collections, Steven and Dorothea Green Library, Florida International University, Miami
leg.	*legajo*
MP	Mapas y Planos, SD
PC	Papeles Procedentes de Cuba, AGI
PN	Protocolos Notariales, Escribania de Gobierno, Archivo Historico Provincial de Pinar del Río, Cuba
Población	*AGI, AHN, Cuba, Población*, 1766–1851, vol. 119, LMC
RC	Real Consulado y Junta de Fomento, ANC
SC	Spanish Consulado at Norfolk, VHS
SD	Audiencia de Santo Domingo, AGI
SGU	Secretaría del Despacho de Guerra, AGS
VHS	Virginia Historical Society, Richmond

INTRODUCTION

1. Christopher Columbus, *The Journal of Christopher Columbus (during his First Voyage, 1492–93), and Documents relating to the Voyages of John Cabot and Gaspar Corte Real*, translated with notes and an introduction by Clements R. Markham (London: Hakluyt Society, 1893), 61, 71; Fernando Ortiz, *Cuban Counterpoint: Tobacco and Sugar*, translated by Harriet de Onís, with an introduction by Bronislaw Malinowski, prologue by Herminio Portell Vilá, and new introduction by Fernando Coronil (Durham, NC: Duke University Press, 1995 [1947]), 105–8, 155; Marcy Norton, *Sacred Gifts, Profane Pleasures: A History of Tobacco and Chocolate in the Atlantic* (Ithaca, NY: Cornell University Press, 2008), 1, 45.
2. Christopher Columbus, *The "Diario" of Christopher Columbus's First Voyage to America, 1492–1493*, abstracted by Fray Bartolomé de las Casas, and transcribed and translated into English, with notes and a concordance of the Spanish, by Oliver Dunn and James E. Kelley Jr. (Norman: University of Oklahoma Press, 1989), 79, 83, 87, 111, 145, 221, 239, 301–2.
3. Georg Friederici, *Der Charakter der Entdeckung und Eroberung Amerikas durch die Europäer*, vol. 1 (Stuttgart: Perthes, 1925), 404, cited in Urs Bitterli, *Cultures in Conflict: Encounters between European and Non-European Cultures, 1492–1800* (Stanford, CA: Stanford University Press, 1989), 75; J. A. P. Jones, *Europe, 1500–1600* (Surrey, UK: Thomas Nelson, 1997), 298. Although gold is emphasized here, Spanish and other European explorers were interested in mineral wealth of any kind, including silver, which proved to be far more abundant. In "The Sustained Debts of Philip II: A Reconstruction of Castile's Fiscal Position, 1566–1596," however, Mauricio Drelichman and Hans-Joachim Voth argue that Philip's were "sustainable," noting that even the defeat of the Spanish Armada caused "mild" deterioration in its economic position. Instead, they contend, "Castile's finances mainly suffered large, temporary shocks as a result of military events." *Journal of Economic History* 70, no. 4 (December 2010): 814.
4. Joseph C. Winter, "Botanical Description of the North American Tobacco Species," in *Tobacco Use by Native North Americans: Sacred Smoke and Silent Killer*, edited by Joseph C. Winter (Norman: University of Oklahoma Press, 2000), 89–90; W. A. Brennan, *Tobacco Leaves: Being a Book of Facts for Smokers* (Menasha, WI: Collegiate Press, 1915), 14.
5. David Armitage, "Three Concepts of Atlantic History," in *The British Atlantic World, 1500- 1800*, edited by David Armitage and Michael J. Braddick (New York: Palgrave Macmillan, 2002), 15.
6. Johannes Wilbert, *Tobacco and Shamanism in South America* (New Haven, CT: Yale University Press, 1987), xv.
7. Ramón de la Sagra, *Historia fisica, politica y natural de la isla de Cuba* (Paris: A. Bertrand, 1839), 284. Where not otherwise indicated, translations from Spanish-language sources are mine.
8. José Rivero Muñiz, *Tabaco: Su historia en Cuba*, vols. 1–2 (Havana: Instituto de Historia, 1964–1965); Ramiro Guerra y Sánchez et al., *Historia de la nación cubana*, vols. 2–3 (Havana: Editorial Historia de la Nación Cubana, 1952); Ramiro Guerra y Sánchez et al., *Ilustración libertad de comercio (desde 1790 hasta 1837)*, vol. 3 of *Historia de la nación cubana* (Havana: Editorial Historia de la Nación Cubana, 1952); Manuel

Moreno Fraginals, El *ingenio: Complejo económico social cubano del azúcar* (Havana: Editorial de Ciencias Sociales, 1978 [1964]).

9. Gaspar Jorge García Galló, *Biografía del tabaco habano* (Havana: Universidad Central de Las Villas, Departamento de Relaciones Culturales, 1959), 39.
10. *Tazmias* were in use as early as the 1750s. See, for example, Francisco Cagigal de la Vega to Capitán General, 12 November 1750, Santiago de Cuba, order 158, *legajo* (bundle; hereafter leg.) 5, Correspondencia de los Capitanes Generales (hereafter CG), Archivo Nacional de Cuba (hereafter ANC), Havana; Alonso Arcos y Moreno to Cagigal de la Vega, 27 March 1752, Santiago de Cuba, order 74, leg. 6, CG; Cagigal de la Vega to unknown, 16 March 1756, order 164, leg. 7, CG. Most of the *tazmias* used in this study date from 1768 to 1783. In legs. 1003 and 1060 of the Intendente General de Hacienda (hereafter IGH), ANC, there are several references to *tazmias* after 1783, but only two such documents were found among the Spanish and Cuban *legajos* consulted. See *tazmia*, Partidos of Rio de Agabama, Rio de Ay, Caracucey, Yguanojo, Animado, and Caunado, 7 June 1789, leg. 1313, Papeles Procedentes de Cuba (hereafter PC), Archivo General de Indias (hereafter AGI), Seville; "Autos sobre el mal uso que hacen los vegueros de Chambas y Tatibunico de sus cosechas de tabaco," 22 November 1798, San Juan de Remedios, no. 7, leg. 1003, IGH.
11. See, for example, Sherry Johnson, *The Social Transformation of Eighteenth-Century Cuba* (Gainesville: University Press of Florida, 2001).
12. Within Spain's colonial bureaucratic system, a captain general provided military leadership, while a governor represented civil authority. However, in the case of Cuba, the island's highest political authority was the "captain general and governor," a title held by one individual. For the purposes of this study, the titles will be used interchangeably when referring to Cuba's colonial leaders.
13. Rudyard Kipling, *The Collected Poems of Rudyard Kipling* (Ware, UK: Wordsworth Editions, 1994), 52.

CHAPTER 1

1. José de Gálvez, license for Antonio Aguilar de Cela, 9 December 1777, Madrid, folio 1, leg. 5524, Casa de Contratación (hereafter Contratación), AGI; Samuel Hazard, *Cuba with Pen and Pencil* (Miami: Editorial Cubana, 1989 [1871]), 215; John George F. Wurdemann, *Notes on Cuba* (Boston: James Munroe, 1844), 32–34, 57–59, 220–25, 228–35; Robert M. Levine, *Cuba in the 1850s: Through the Lens of Charles DeForest Fredricks* (Tampa: University of South Florida Press, 1990), 37; Sherry Johnson, *Social Transformation*, 29; Leví Marrero, *Azucar, ilustración y conciencia (1763–1868) (III)*, vol. 11 of *Cuba: Economía y sociedad* (Madrid: Editorial Playor, 1984), 8. The Francisco Miranda who accompanied Antonio Aguilar de Cela is not the Venezuelan revolutionary who fought against the Spanish and later died in a Spanish prison.
2. Antonio Aguilar de Cela to Diego José Navarro, 4 August 1778, Havana, leg. 1243, PC. Aguilar likely entered the San Carlos de la Caridad, a military hospital founded during the 1761 yellow fever epidemic, or the Nuestra Señora de Pilar hospital that served the king's slaves and became associated with the nearby military Hospital San Ambrosio. Both were located close to the royal tobacco factory.

See Eduarda Ancheta Niebla, "La Orden de San Juan de Dios y el Hospital de San Felipe y Santiago" (PDF version at Nicanor Aniorte Hernández, *Página personal sobre la Formación Postgrado en Enfermería*, *www.aniorte-nic.net*), 10–11 (first published at *Enfermería Avanza*, 19 April 2012, *enfeps.blogspot.com.es*).

3. Jean Stubbs, "El Habano: The Global Luxury Smoke" (Commodities of Empire Working Paper No. 20, Open University Ferguson Centre of African and Asian Studies [Milton Keynes, UK] and Institute of the Americas [London], September 2012), *www.open.ac.uk*.
4. "Articles of Capitulation," *Providence (RI) Gazette* 1, no. 3 (6 November 1762): 1; Celia María Parcero Torre, *La pérdida de la Habana y las reformas borbónicas en Cuba, 1760–1773* (Valladolid, Spain: Junta de Castilla y León, Consejería de Educación y Cultura, 1998) provides a detailed analysis of the loss of Havana.
5. Winter, "Botanical Description," 114; Joseph C. Winter, "From Earth Mother to Snake Woman: The Role of Tobacco in the Evolution of Native American Religious Organization," in Winter, *Tobacco Use*, 268–69.
6. Jordan Goodman, *Tobacco in History: The Cultures of Dependence* (London: Routledge, 1993), 20–24, 27–30, 33; Ortiz, *Cuban Counterpoint*, 105, 155; Peter C. Mancall, "Tales Tobacco Told in Sixteenth-Century Europe," *Environmental History* 9, no. 4 (October 2004): 650–51; Wilbert, *Tobacco and Shamanism*, 9–132; Bernal Díaz, *The Conquest of New Spain*, translated and with an introduction by J. M. Cohen (New York: Penguin, 1988 [1963]), 227.
7. Norton, *Sacred Gifts*, 56; Goodman, *Tobacco in History*, 51–52.
8. Norton, *Sacred Gifts*, 53–59, 105, 108, 127; Goodman, *Tobacco in History*, 44–47; Anne Charlton, "Medicinal Uses of Tobacco in History," *Journal of the Royal Society of London* 97 (June 2004): 292–94. Sugar initially was also touted as having medicinal benefits. See Sidney W. Mintz, *Sweetness and Power: The Place of Sugar in Modern History* (New York: Penguin, 1986), 96.
9. Norton, *Sacred Gifts*, 56.
10. Ibid., 119; James I, *A Counterblaste to Tobacco* (London: 1604); R. T. Ravenholt, "Tobacco's Global Death March," *Population and Development Review* 16, no. 2 (June 1990): 216; Jacob Sullum, *For Your Own Good: The Anti-Smoking Crusade and the Tyranny of Public Health* (New York: Simon and Schuster, 1999), 21. One scholar argues that James I's decision to publish *Counterblaste* was not solely due to his opposition of tobacco, but rather because its trade "posed a number of threats to James's own sense of power." See Susan Campbell Anderson, "A Matter of Authority: James I and the Tobacco War," *Comitatus: A Journal of Medieval and Renaissance Studies* 29, no. 1 (1998): 136–63.
11. Ortiz, *Cuban Counterpoint*, 224, 234–36.
12. Rudi Matthee, *The Pursuit of Pleasure: Drugs and Stimulants in Iranian History, 1500–1900* (Princeton, NJ: Princeton University Press, 2005), 135–38; W. F. Axton, *Tobacco and Kentucky* (Lexington: University Press of Kentucky, 1975), 25.
13. Ravenholt, "Tobacco's Global Death March," 220; Rae Duncan Laurenson, "The Emperor Who Smoked a Pipe," *Journal of Laryngology and Otology* 109, no. 1 (January 1995): 1–4. Laurenson notes that Frederick III likely smoked for thirty years prior to his death.
14. Ravenholt, "Tobacco's Global Death March," 220, 228. A 1602 publication, *Work for Chimney-Sweepers or A Warning for Tobacconists*, was the first to cite the health risks

of tobacco, arguing that tobacco worsened some diseases, was addictive, and posed health risks to other organs, including the heart and skin, among its other findings. See Anne Charlton, "Tobacco or Health 1602: An Elizabethan Doctor Speaks," *Health Education Research* 20, no. 1 (2005): 101–11.

15. Leví Marrero, *El siglo XVII (II)*, vol. 4 of *Cuba: Economía y sociedad* (Madrid: Editorial Playor, 1975), 38–40; cited in H. E. Friedländer, *Historia económica de Cuba* (Havana: Jesus Montero, 1944), 14:65. Contemporary English writers frequently spoke out against the popularity of Spanish tobacco, reflecting a period distrust of all foreigners and their products, but particularly the Spanish who were a frequent enemy. See Susan Campbell Anderson, "Matter of Authority," 154.
16. Norton, *Sacred Gifts*, 142–43; Marrero, *Cuba*, 4:38–40; Virginius Dabney, *Virginia: The New Dominion* (Charlottesville: University Press of Virginia, 1996 [1971]), 25; Goodman, *Tobacco in History*, 148–49, 180; Alan Kulikoff, *Tobacco and Slaves: The Development of Southern Cultures in the Chesapeake, 1680–1800* (Chapel Hill: University of North Carolina Press, 1986), 32; Axton, *Tobacco and Kentucky*, 25. Axton credits Cuba for Virginia's success in tobacco cultivation, arguing that *Nicotiana tabacum* seed was smuggled out of Cuba and brought to the British colony to replace the inferior local tobacco, *Nicotiana rusticum*, grown by Virginia's Native Americans.
17. Carole Shammas, *The Pre-industrial Consumer in England and America* (Oxford: Oxford University Press, 1990), 40.
18. Norton, *Sacred Gifts*, 143; Marrero, *Cuba*, 4:38–39. Virginia also experimented with sugar cane cultivation, although unsuccessfully. Introduced as early as 1619, the same year that African slaves were brought to the English colony, sugar cane did not thrive in the area. See Mintz, *Sweetness and Power*, 37.
19. José M. Rodríguez Gordillo, *La diffusion del Tabaco en España: Diez estudios* (Seville: Universidad de Sevilla, Fundación Altadis, 2002), 25–28; Agustín González Enciso, "Los usos financieros del tabaco; o bien, para qué servían los dineros obtenidos con la renta del tabaco," in *El mercado del tabaco en España durante el siglo XVIII: Fiscalidad y consumo*, edited by Santiago de Luxán Meléndez, Sergio Solbes Ferri, and Juan José Laforet (Las Palmas, Spain: Universidad de Las Palmas de Gran Canaria, Servicio de Publicaciones, 2000) 40; José M. Rodríguez Gordillo, *Un archivo para la historia del tabaco* (Madrid: Jacaryan, 1984), 23–49.
20. Ravenholt, "Tobacco's Global Death March," 218–19; Walter J. Psoter and Douglas E. Morse, "Annual Per Capita Apparent Consumption of Tobacco Products in the United States: 1900–1990," *Preventative Medicine* 32 (2001): 7.
21. "Tobacco Crops in Cuba: Worth More to the Island than Sugar, Coffee," *Chicago Daily Tribune*, 21 November 1898, 9.
22. "The Texas State Fair," *Dallas Morning News*, 2 November 1892, 2.
23. "The Marysville District Fair, Marysville, Wednesday, September 5, 1860," *San Francisco Daily Evening Bulletin*, 7 September 1860, 3.
24. "Mistakes of Smokers: Cigars that Pass for Havana Not Made of the Real Abajo Tobacco," *Columbus (OH) Daily Enquirer-Sun*, 20 March 1883, 1.
25. Antonio Garcia Cossio to Marqués de Robledo, 22 October 1779, Cádiz, leg. 610, Archivo de la Fábrica de Tabacos de Sevilla (hereafter AFTS), Seville; unknown to Administeres Generales de la Renta del Tavaco del Reino, 15 May 1781, Seville, leg. 612, AFTS; Miguel Cayetano Soler to Sr. Director de la Renta de Tabaco de la Isla de Cuba, 1 April 1804, Aranjuez, leg. 2018, Audiencia de Santo Domingo

(hereafter SD), AGI. The transatlantic component of the tobacco trade is detailed in Chapter 7.

26. Goodman, *Tobacco in History*, 47; Mintz, *Sweetness and Power*, 45; Norton, *Sacred Gifts*, 143, 161.
27. Quoted in Norton, *Sacred Gifts*, 161.
28. Stanley Gray and V. J. Wyckoff, "The International Tobacco Trade in the Seventeenth Century," *Southern Economic Journal* 7, no. 1 (July 1940): 12.
29. Richard R. Madden, *La isla de Cuba: Sus recuerdos, progresos y perspectivas*, translated by Sarah Méndez Capote (Havana: Consejo Nacional de Cultura, 1964 [1849]), 189; Joseph Judson Dimock, *Impressions of Cuba in the Nineteenth Century: The Travel Diary of Joseph J. Dimock*, edited by Louis A. Pérez Jr. (Wilmington, DE: Scholarly Resources, 1998), 93.
30. Richard Henry Dana Jr., *To Cuba and Back*, edited with an introduction by C. Harvey Gardiner (Carbondale: Southern Illinois University Press, 1966 [1859]), 102; Luis Martínez-Fernández, *Fighting Slavery in the Caribbean: The Life and Times of a British Family in Nineteenth-Century Havana* (Armonk, NY: M. E. Sharpe, 1998), 20, 68.
31. Norton, *Sacred Gifts*, 22–23.
32. Matthew Hilton, "Leisure, Politics, and the Consumption of Tobacco in Britain since the Nineteenth Century," in *Histories of Leisure*, edited by Rudy Koshar (Oxford: Berg, 2002), 322.
33. Paul R. Johnson, *The Economics of the Tobacco Industry* (New York: Praeger, 1984), 7.
34. Peter Borsay, *A History of Leisure: The British Experience since 1500* (New York: Palgrave Macmillan, 2006), 2–3.
35. Peter Burke, "Viewpoint: The Invention of Leisure in Early Modern Europe," *Past and Present* 146 (February 1995): 136–50; Borsay, *History of Leisure*, 9, 195; Hilton, "Leisure, Politics," 319–21.
36. David T. Courtwright, *Forces of Habit: Drugs and the Making of the Modern World* (Cambridge, MA: Harvard University Press, 2001), 2, 59.
37. Charlton, "Tobacco or Health 1602," 105.
38. *Tenth Report of the Commissioners of Enquiry into the Revenue Arising in Ireland*, British Parliamentary Papers, House of Commons (London: HMSO, 1824), x, quoted in Andy Bielenberg and David Johnson, "Production and Consumption of Tobacco in Ireland, 1800–1914," *Irish Economic and Social History* 25 (1998): 2. Mintz notes that sugar offered similar hunger-fighting properties (*Sweetness and Power*, 134).
39. Burke, "Viewpoint," 136–50; Eleanor Elsner, *Romantic Spain: The Land of Mantillas and Matadors* (London: Butterworth, 1924), 144; Hilton, "Leisure," 320–32.
40. Charles Davillier, *Spain*, translated by J. Thomson (New York: Scribner, Welford, and Armstrong, 1876 [1874]), 274, quoted in M. Elizabeth Boone, *Vistas de España: American Views of Art and Life in Spain, 1860–1914* (New Haven, CT: Yale University Press, 2007), 128.
41. Boone, *Vistas de España*, 128.

CHAPTER 2

1. Josef Pablo Valiente, letter and testimony, 4 July 1799, Havana, no. 72, leg. 2000, SD; Pedro Antonio Gamon to Juan José de la Hoz, 24 January 1804, Real Factoría de Tabacos, Havana, no. 47, leg. 1060, IGH; Ignacio González Cadura, 3 March 1810, leg. 2000, SD; "Demostración hecha por la Contaduria . . . tabaco rama Vir-

ginia. . . ." (table), 15 November 1810, Havana, leg. 2001, SD; José de Ymaz et al., "Condiciones con que se ha verificado la contrata general de tabaco hoja Habana, aprobada por SM en real orden de cuatro de agosto ultimo en favor de don Angel Moreno e hijos, del comercio de esta corte," 28 October 1833, Madrid, Ultramar, AGI; [writing unclear], Fábrica de Tabacos de Sevilla, Contaduria, Clases de tabaco para las Rentas de Tabaco de La Palma, 21 January 1868, Seville, leg. 2282, AFTS; [writing unclear], Fábrica de Tabacos de Sevilla, Contaduria, Clases de tabaco para las Rentas de Tabaco de Huelva, 21 January 1868, Seville, leg. 2282, AFTS; [*writing unclear*], Fábrica de Tabacos de Sevilla, Contaduria, Clases de tabaco para las Rentas de Tabaco de Cordoba, 21 January 1868, Seville, leg. 2282, AFTS.

2. T. H. Breen, *Tobacco Culture: The Mentality of the Great Tidewater Planters on the Eve of the Revolution*, 2nd ed. (Princeton, NJ: Princeton University Press, 2001), 44; Goodman, *Tobacco in History*, 13.
3. Brennan, *Tobacco Leaves*, 21–22.
4. Quoted in Ortiz, *Cuban Counterpoint*, 27.
5. Breen, *Tobacco Culture*, 44–55; Ortiz, *Cuban Counterpoint*, 27–33, 56; William Morton, "Diaries," 1846 and 1847, Mss2 M8464 b, William Morton Papers, Virginia Historical Society (hereafter VHS), Richmond.
6. C [*illegible*] Arguelles to Superintendente General de la Factoría de Tavacos de la Havana, 28 June 1811, Cádiz, *expediente* (record; hereafter exp.) 22, leg. 973, IGH; Juan Antonio de Unzueta, "Aviso al publico," 17 September 1811, exp. 22, leg. 973, IGH.
7. Marqués de C [*illegible*] to Sr. Ministro de la Guerra y Ultramar, 15 April 1859, exp. 2, leg. 50, Ultramar, Archivo Histórico Nacional (hereafter AHN), Madrid; Lino Lopez Mendez, *Manual del veguero venezolano: Cultivo del tabaco segun los mejores metodos empleados en Cuba y adaptados a la practica en Venezuela* (Caracas: Imprenta al Vapor de "La Opinion Nacional," 1883); Pedro Gamon to Luis de Viguri, 18 August 1802, Havana, leg. 235, Ultramar, AGI.
8. Miguel de Múzquiz to Marqués de la Torre, 24 January 1776, El Prado, folios 896–97, leg. 1227, PC. There are at least two other references to copying Brazilian tobacco: see Múzquiz to de la Torre, 27 December 1775, Madrid, folios 892–93, leg. 1227, PC; and Mateo de Echavarría y Elguerra to de la Torre, 11 November 1776, Santiago de Cuba, leg. 1156, PC.
9. Barbara Hahn, *Making Tobacco Bright: Creating an American Commodity, 1617–1937* (Baltimore: Johns Hopkins University Press, 2011), 2.
10. Un Amigo del Pais, *Memoria sobre el tabaco de la isla de Cuba, en la que se indican algunas mejoras de que es susceptible su cultivo y preparacion* (Havana: Imprenta de Antonio María Davila, 1852), 5; Martín Fernández Miñano, "Cultura de tabaco en la isla de Cuba," 1788, leg. 2002, SD; De la Sagra, *Historia fisica, politica y natural*, 288; Peter Minor, *Notes on the Cultivation and Management of Tobacco, from the Plant Bed to the Prize; According to the Most Approved Practices in Albemarle, and the Adjacent Counties in Virginia* (Baltimore: J. Robinson, 1822), 3; J. F. Nobiatur, *El veguero practico y cientifico, o sea Cultivo perfeccionado del tabaco con el méthodo de preparar convenientemente la hoja, para sus más ventajosa elaboracíon, todo con arreglo á los adelantados modernos* (Havana: El Telegrafo, 1874), 5–7.
11. Timothy Silver, *A New Face on the Countryside: Indians, Colonists, and Slaves in South Atlantic Forests, 1500–1800* (Cambridge: Cambridge University Press, 1990), 59–62.

"Wasteful and destructive" is from Hu Maxwell, "The Use and Abuse of the Forests by the Virginia Indians," *William and Mary Quarterly College Magazine*, 2nd ser. 19 (October 1910), 86, quoted in Silver, *New Face*, 59.

12. D. W. Meining, *Atlantic America, 1492–1800*, vol. 1 of *The Shaping of America: A Geographical Perspective on 500 Years of History* (New Haven, CT: Yale University Press, 1986), 154; Silver, *New Face*, 163; Brennan, *Tobacco Leaves*, 43; Kulikoff, *Tobacco and Slaves*, 47.
13. David O. Percy, "Ax or Plow? Significant Colonial Landscape Alteration Rates in the Maryland and Virginia Tidewater," *Agricultural History* 66, no. 2 (Spring 1992): 69. Percy notes that at least two tobacco crops could be grown before another crop was substituted. The choice was usually corn, as its deep taproots accessed nutrients unavailable to the surface-feeding tobacco plants (71–72).
14. *Cabildo* decree, city of Trinidad, 25 April 1766, leg. 1087, PC; Minor, *Notes*, 7–8. As noted earlier in this chapter, this view was unique, as most Virginia planters felt that tobacco exhausted the soil.
15. Quoted in Susan Deans-Smith, *Bureaucrats, Planters, and Workers: The Making of the Tobacco Monopoly in Bourbon Mexico* (Austin: University of Texas Press, 1992), 116.
16. Minor, *Notes*, 14; Un Amigo del Pais, *Memoria sobre el tabaco*, 7; "Methodo de sembrar y cosechar el Tabaco en la Ysla de Cuba," n.d., leg. 2002, SD. Some locations, such as the Dominican Republic, skipped this step altogether, broadcasting tobacco directly into the fields instead of using seedbeds. See Michiel Baud, *Peasants and Tobacco in the Dominican Republic, 1870–1930* (Knoxville: University of Tennessee Press, 1995), 58.
17. Nobiatur, *El veguero*, 9; Minor, *Notes*. In contrast, Venezuelan farmers sowed their seedbeds from November to January, and could plant a second seedbed in mid-June (Lopez Mendez, *Manual del veguero venezolano*, 20). T. H. Breen references an 1800 publication by William Tatham, *An Historical and Practical Essay on the Culture and Commerce of Tobacco* (reprinted in *William Tatham and the Culture of Tobacco*, by G. Melvin Herndon (Coral Gables, FL: University of Miami, 1969), 118–19, that places the appropriate date for establishing the seedbed in Virginia as late December or early January.
18. T. C. Tso, *Physiology and Biochemistry of Tobacco Plants* (Stroudsburg, PA: Dowden, Hutchinson and Ross, 1972), 11.
19. Fernández Miñano, "Cultura de tabaco."
20. Ibid.; Nobiatur, *El veguero*, 8–10; Minor, *Notes*, 4.
21. Fernández Miñano, "Cultura de tabaco."
22. Minor, *Notes*, 4; George B. Lucas, *Diseases of Tobacco* (New York: Scarecrow Press, 1958), 453.
23. Francisco Gama to Antonio María de Bucarely, 4 November 1768, Güines, leg. 1090, PC.
24. Nobiatur, *El veguero*, 12, 13; Minor, *Notes*, 4; Lopez Mendez, *Manual del veguero venezolano*, 23, 27; "Methodo de sembrar"; "Sobre el ramo del tabaco y medios de reestablecerlo," n.d., leg. 250, Ultramar, AGI; Antonio María de Paula Arias, *El veguero de Vuelta Abajo: Apuntes sobre el cultivo del tabaco* (Pinar del Río, Cuba: Imprenta de Miguel Vives, 1887), 11; Un Amigo del Pais, *Memoria sobre el tabaco*, 49; Goodman, *Tobacco in History*, 171.
25. Fernández Miñano, "Cultura de tabaco"; Un Amigo del Pais, *Memoria sobre el*

tabaco, 25; Nobiatur, *El veguero*, 16, 17; "Methodo de sembrar"; Lopez Mendez, *Manual del veguero venezolano*, 34, 36; Minor, *Notes*, 6.

26. Washington, quoted in Breen, *Tobacco Mentality*, 81.
27. Breen, *Tobacco Mentality*, 55, 81, 181, 205–6. The Jefferson quote appears on page 206.
28. In 1774, a new administrative entity was created in the far western part of Cuba. Emeterio S. Santovenia notes that over time, the area was called by various names: Filipina, Filipinas, and Nueva Filipinas, with Nueva Filipinas eventually winning out. The name Pinar del Río was given to both the town and the *partido* (district) in which it was located. Today, the western part of the island is known as Pinar del Río Province. Emeterio S. Santovenia, *Pinar del Río* (Mexico: Fondo de Cultura Economica, 1946), 93–94.
29. Alonso Arcos y Moreno to Francisco Cagigal de la Vega, 27 March 1752, Santiago de Cuba, exp. 74, leg. 6, ANC; Sherry Johnson, *Climate and Catastrophe in Cuba and the Atlantic World in the Age of Revolution* (Chapel Hill: University of North Carolina Press, 2011), 203–6; Julián de Arriaga to Antonio María de Bucarely, 18 July 1769, Madrid, exp. 24, leg. 1134, PC; Joaquin [*illegible*] and Blas de Noriega to unknown, 3 May 1792, leg. 1461, PC; Marqués de Guisa to Juan de Mecolaeta, 14 May 1801, Bayamo, leg. 235, Ultramar, AGI. Johnson notes that periods of drought usually were followed by active hurricane cycles.
30. "Cuyaguateje" appears as "Cuiaguatexe" in a number of period documents.
31. Antonio Fernández to Marqués de la Torre, 20 June 1775, Filipinas, leg. 1180, PC; Fernández to de la Torre, 25 June 1775, Filipinas, leg. 1180, PC.
32. Ramón Lloret to Diego José Navarro, 12 April 1779, Filipinas, leg. 1261, PC.
33. Julio Le Riverend Brusone, *La Habana, espacio y vida* (Madrid: Editorial MAPFRE, 1992), 23; Fred A. Carlson, *Geography of Latin America*, 3rd ed. (New York: Prentice-Hall, 1952), 443; Louis A. Pérez Jr., *Winds of Change: Hurricanes and the Transformation of Nineteenth-Century Cuba* (Chapel Hill: University of North Carolina Press, 2001), 31–32; Adolfo Dollero, *Cultura cubana: La provincia de Pinar del Río y su evolucíon* (Havana: Seoane, Fernández, 1921), 33; Sherry Johnson, *Climate and Catastrophe*; Miguel Guerrero to Marqués de la Torre, 19 November 1774, Partido de Guane al Sur, leg. 1180, PC; Francisco Gelabert, "Relación del reconocimiento. . . ." 12 July 1776, Havana, typescript copy in *Documentos sobre Cuba 1740–1806*, vol. 27 (hereafter DSC), Leví Marrero Collection (hereafter LMC), Special Collections, Steven and Dorothea Green Library, Florida International University, Miami.
34. Miguel Guerrero to Marqués de la Torre, 19 November 1774, Partido de Guane al Sur, leg. 1180, PC; José Perez to de la Torre, 21 November 1774, Pinar del Río, leg. 1180, PC; Domingo Duarte to Perez, 19 November 1774, Consolación del Sur, leg. 1180, PC.
35. Jorge Macle Cruz, "La tormenta de San Francisco de Borja," *Boletín del Archivo Nacional* 12 (2000): 108; letter to the editor, *Diario de la Marina* (Havana), 8 October 1846, 3, quoted in Pérez, *Winds of Change*, 69.
36. El Apoderado de la Real Compañía to Capitán General, 1 August 1755, Santiago de Cuba, order 407, leg. 7, CG.
37. Joseph de [*illegible*], "Estado que manifiesta las ruinas que ocasiono el urucán. . . ." 25 October 1768, Guanabacoa, leg. 1090, PC.

38. Francisco Gama to Antonio María de Bucarely, 4 November 1768, Güines, leg. 1090, PC.
39. Vicente de Mesa to Marqués de la Torre, 13 November 1774, Managua, leg. 1193, PC.
40. Manuel Ramos to Marqués de la Torre, 10 November 1774, Río Blanco, leg. 1194, PC; Francisco de la Cruz Valiente to de la Torre, 14 November 1774, Govea, leg. 1196, PC; Sebastian de la Cruz to de la Torre, 17 November 1774, Güines, leg. 1193, PC; de la Cruz to de la Torre, 19 November 1774, Güines, leg. 1193, PC.
41. Mateo de Echavarría y Elguerra to Diego José Navarro, 9 December 1777, Santiago de Cuba, leg. 1231, PC.
42. Sherry Johnson, *Climate and Catastrophe*.
43. *Relación*, 25 April 1795, no. 41, leg. 2612, AGI.
44. Richard Bland to Thomas Adams, 1 August 1771, quoted in Dennis B. Blanton, Michael Chenoweth, and Cary J. Mock, "The Great Flood of 1771: An Explanation of Natural Causes and Social Effects," in *Historical Climate Variability and Impacts in North America*, edited by Lesley-Ann Dupigny-Giroux and Cary J. Mock (Dordrecht, Netherlands: Springer, 2009), 4–5.
45. "Williamsburg, in Virginia, June 1," *Connecticut Journal* (New Haven), no. 192 (21 June 1771): 2; "Williamsburg (Virginia) June 6," *Providence Gazette and Country Journal* (Providence, RI) 8, no. 390: 102; "Further Accounts of the Damage Done by the Floods in Virginia, Williamsburg, in Virginia, June 6," *Boston Evening-Post*, no. 1866: 2.
46. St. George Tucker, *Hansford: A Tale of Bacon's Rebellion* (Richmond, VA: George M. West, 1857), 15.
47. Tatham, *Historical and Practical Essay*, 22.
48. Joseph Clarke Robert, *The Tobacco Kingdom: Plantation, Market, and Factory in Virginia and North Carolina, 1800–1860* (Gloucester, MA: Peter Smith, 1965), 37.
49. Minor, *Notes*, 8.
50. Ibid.; "Methodo de sembrar"; Un Amigo del Pais, *Memoria sobre el tabaco*, 27; Lopez Mendez, *Manual del veguero venezolano*, 42.
51. Breen, *Tobacco Culture*, 49–50, 63–68; Minor, *Notes*, 8.
52. Fernández Miñano, "Cultura de Tabaco"; Nobiatur, *El veguero*, 19; "Methodo de sembrar"; Minor, *Notes*, 9.
53. Fernández Miñano, "Cultura de Tabaco."
54. Minor, *Notes*, 9.
55. Fernández Miñano, "Cultura de tabaco"; Nobiatur, *El veguero*, 19–20; Minor, *Notes*, 9–11.
56. Carlson, *Geography of Latin America*, 441–42; Nobiatur, *El veguero*, 20–25.
57. Gobernador de Cuba to Francisco Cagigal de la Vega, 2 September 1752, Santiago de Cuba, order 195, leg. 6, CG; Cagigal de la Vega to Alonso Arcos y Moreno, 25 October 1752, Havana, order 299, leg. 6, CG; Joaquín de Sola to Marqués de la Torre, 6 July 1776, Puerto Príncipe, leg. 165, PC.
58. Ramón Vuelta Flores to Antonio María de Bucarely, 20 August 1769, Puerto Príncipe, leg. 1074, PC. The monopoly's scramble for funds to pay the *vegueros* is discussed in detail in later chapters, and is reflected in period correspondence such as Francisco Cagigal de la Vega to Lorenzo de Madariaga, 6 September 1754, Havana, leg. 6, CG.

59. Marqués de la Torre to Alcaldes de San Felipe y Santiago, 15 November 1775, Havana, leg. 1167, PC; Domingo de la Cruz to de la Torre, 21 November 1775, Santiago, leg. 1167, PC.
60. Ambrocio Pérez Baes to Marqués de la Torre, 8 December 1772, Partido de Macuriges, leg. 1193, PC.
61. Marqués de la Torre to Alcaldes de la Ciudad de Santiago, 29 March 1776, Havana, leg. 1167, PC; de la Torre to Alcaldes de San Felipe y Santiago de Bejucal, 29 March 1776, Havana, leg. 1167, PC; Mathias de la Cruz to de la Torre, 26 April 1776, Bejucal, leg. 1167, PC. The mule pack drivers were Juan González, Francisco Alonzo, Antonio Beltrán, Felix Castellaño, Juan Farias, Isidoro Garcia, Antonio Gómez, Francisco González, Pedro Hernández, Francisco Isquierdo, José de Leon, José Marrero, Juan Padron, Esteban Pastrana, Blas Rivero, and Juan Tavarez.
62. Directores de la Renta to Julián de Arriaga, 24 December 1757, Havana, leg. 1002, Ultramar, AGI; Luis de Viguri to Miguel Cayetano Soler, 8 February 1801, Havana, letter 89, leg. 2000, SD.
63. Miguel Ibáñez Cuevas to Antonio María de Bucarely, 26 June 1766, Bayamo, leg. 1080, PC; Juan Gemir y Lleonart to Bucarely, 12 June 1767, Bayamo, leg. 1080, PC; Mateo de Echavarría y Elguerra to Bucarely, 25 August 1770, Santiago de Cuba, leg. 1080, PC; Juan Antonio Royo, "Relación de las Embarcaciones que han entrado y salido por este Puerto de la Boca del Morro de Santiago de Cuba desde primo de Henro [*sic*] de 1774 hasta fin del dho," 4 February 1774, Castillo del Morro, Santiago de Cuba, leg. 1141, PC; Gobernador de la Plaza de Cuba to Marqués de la Torre, 22 September 1775, Santiago de Cuba, doc. 37, leg. 1142, PC. For tobacco from Puerto Príncipe and Holguín, see Bucarely to Ibáñes Cuebas, 8 September 1768, Havana, leg. 1077, PC; Bucarely to Ibáñez Cuebas, 20 November 1768, doc. 86, leg. 1077, PC; Ramón Vuelta Flores to Bucarely, 20 September 1769, Puerto Príncipe, leg. 1074, PC; Vuelta Flores to Bucarely, 26 October 1769, Puerto Príncipe, leg. 1074, PC; Antonio de Luque to de la Torre, 25 October 1772, Holguín, doc. 2, leg. 1144, PC; Luque to de la Torre, 1 March 1773, Holguín, doc. 3, leg. 1144, PC; and Luque to de la Torre, 14 October 1773, Holguín, doc. 4, leg. 1144, PC.
64. Sherry Johnson, "'*La Guerra Contra los Habitantes de los Arrabales*': Changing Patterns of Land Use and Land Tenancy in and around Havana, 1763–1800," *Hispanic American Historical Review* 77, no. 2 (1997): 189, 192.
65. Ibid., 192; Hazard, *Cuba*, 214–15.
66. Alexander von Humboldt, *The Island of Cuba*, translated by J. S. Thrasher (New York: Derby and Jackson, 1856), 105; Hazard, *Cuba*, 215; Wurdemann, *Notes on Cuba*, 228; Sherry Johnson, *Social Transformation*, 29; Marrero, *Cuba*, 11:8. Johnson notes that the original crane was destroyed during the British occupation of Havana (1762–1763).
67. Luis Prieto to Diego José Navarro, 24 September 1778, Havana, leg. 1247, PC.
68. Ibid.; Leví Marrero, *Del monopolio hacia la libertad comercial (1701–1763) (II)*, vol. 7 of *Cuba: Economía y sociedad* (Madrid: Editorial Playor, 1978), 96; John Esaias Warren, *Vagamundo; or the Attaché in Spain: Including a Brief Excursion into the Empire of Morocco* (New York: Scribner, 1851), 150. Today the building houses the Universidad de Sevilla, retaining much of its eighteenth- and nineteenth-century character, which can be seen in features such as the tobacco motifs gracing the main entrance and a large fountain in the central courtyard.

CHAPTER 3

1. Quoted in Manuel Moreno Fraginals, *The Sugarmill: The Socioeconomic Complex of Sugar in Cuba, 1760–1860*, translated by Cedric Belfrage (New York: Monthly Review Press, 1976), 21–22.
2. Joaquín de Sola, *tazmia*, 25 May 1770, Puerto Príncipe, leg. 1085, PC; Ramón José Rodríguez, *tazmia*, 15 May 1770, Matanzas, leg. 1090, PC; José de Aguilar y Blanco, *tazmia*, 20 March 1782, Sancti Spíritus, leg. 1313, PC.
3. Diego José Navarro, "Estracto del padron general. . . ." 31 December 1788, Havana, printed in *Revista de la Biblioteca Nacional José Martí* 29 (September–December 1987): 25; "De los habitantes de la isla de Cuba. . . ." December 1792, box 8, no. 10, José Escoto Collection, Houghton Library, Harvard University, Boston, MA; de la Sagra, *Historia economico*, 119; Robert C. West and John P. Augelli, *Middle America: Its Land and Peoples*, 3rd ed. (Englewood Cliffs, NJ: Prentice Hall, 1989), 130; Juan Gregorio Núñez and José González, "Plan demostrativo de todas las tierras . . . en 9 de enero de 1812," 31 May 1817, Real Factoría de Tabacos, Havana, leg. 2019, SD; Guerra y Sánchez et al., *Historia*, 3:346.
4. Marrero, *Cuba*, 11:60–61; Rebecca J. Scott, *Slave Emancipation in Cuba: The Transition to Free Labor, 1860–1899* (Princeton, NJ: Princeton University Press, 1985), 8, 12. In an earlier volume, Marrero contends that just 4.7 percent of Cuba's rural population resided on *vegas* in 1846. See Leví Marrero, *Azucar, ilustración y conciencia (1763–1868) (I)*, vol. 9 of *Cuba: Economía y sociedad* (Madrid: Editorial Playor, 1983), 207. The calculations of population numbers and corresponding percentages found in volume 11 indicate that approximately 17 percent appears to be more accurate.
5. Moreno Fraginals, *Sugarmill*, 19–20; Francisco Gama, *tazmia*, 14 April 1768, Güines, leg. 1090, PC; Gama, *tazmia*, 15 May 1771, Güines, leg. 1090, PC; Gama, *tazmia*, 8 June 1774, Güines, leg. 1195, PC; Marrero, *Cuba*, 9:217; "Partido de San Julian de los Güines, Resumen gral del numero de ganados de todas especies, que tienen sus Haciendas, y de los Caros y Carretas que hay en cada una de ellas," 26 June 1792, Güines, leg. 1471, PC; Sherry Johnson, *Social Transformation*, 54–55; Manuel Cuadrado Melo, *Obispado de la Habana su historia a traves de los siglos*, vol. 1, part 1A (Havana: np, 1970), 74.
6. Ortiz, *Cuban Counterpoint*, 65.
7. A sampling of works that continue this argument includes Ramiro Guerra y Sánchez et al., *Historia de la nación cubana*, vols. 2–3, especially 3:50, 52–53; Santovenia, *Pinar del Río*, 91; Franklin W. Knight, *Slave Society in Cuba during the Nineteenth Century* (Madison: University of Wisconsin Press, 1970), 65; Moreno Fraginals, Sugarmill, 21–22; and Allan J. Kuethe and G. Douglas Inglis, "Absolutism and the Enlightened Reform: Charles III, the Establishment of the *Alcabala*, and Commercial Reorganization in Cuba," *Past and Present: A Journal of Historical Studies* 109 (November 1985): 134.
8. Abiel Abbot, *Letters Written in the Interior of Cuba, between the Mountains of Arcana, to the East, and of Cusco, to the West, in the Months of February, March, April, and May, 1828* (Boston: Bowles and Dearborn, 1829), 200; Francisco Xavier de Lamadriz, "Resumen general . . . Jurisdicíon de la Nueva Filipina. . . ." 1 July 1787, Pinar del Río, leg. 1407, PC; Ramón Lloret, "Resumen general . . . Jurisdicíon de la Nueba Población de Filipina. . . ." 1 April 1788, Pinar del Río, leg. 1407, PC; Ramón de

Villoro, "Resumen general . . . Jurisdicíon de la Nueba Población de Filipina. . . ." 1 April 1789, Pinar del Río, leg. 1407, PC; Nicolás Hidalgo Gato, *tazmia*, 10 April 1770, San Juan y Martínez, leg. 1094, PC; Lloret, *tazmia*, 1 June 1779, San Juan y Martínez, leg. 1261, PC; Marrero, *Cuba*, 11:61.

9. Tomás de Sierra, *tazmia*, 24 May 1770, Habarro, leg. 1094, PC; Matias Joseph Duarte, *tazmia*, 6 June 1770, San Ildefonso de Guane, leg. 1094, PC; Francisco de Paula Mari, *tazmia*, 2 June 1770, Pinar del Río, leg. 1094, PC; Nicolás Hidalgo Gato, *tazmia*, 10 April 1770, San Juan y Martínez, leg. 1094, PC; Hidalgo Gato, *tazmia*, 22 June 1770, San Juan y Martínez, leg. 1094, PC; Hidalgo Gato, *tazmia*, 27 March 1771, San Juan y Martínez, leg. 1096, PC; Hidalgo Gato, *tazmia*, 10 March 1772, leg. 1180, PC; Ramón Lloret, *tazmia*, 1 June 1779, San Juan y Martínez, leg. 1261, PC. Only the individual *vegueros* listed as owners were included in this calculation.
10. For biographical information, see "Noticia de los Vecinos . . . Partido de Sn Juan y Martines. . . ." n.d., leg. 1704, PC, photocopy found in AGI, AHN, *Cuba, Población, 1766–1851*, vol. 119, LMC (hereafter *Población*); Vicente Blanco, "Pie de lista de la compañía suelta. . . ." 31 July 1779, San Juan, leg. 1269, PC.
11. Nicolás Hidalgo Gato, *tazmia*, 22 June 1770, San Juan y Martínez, leg. 1094, PC; Gato, *tazmia*, 27 March 1771, San Juan y Martínez, leg. 1096, PC; Ramón Lloret, *tazmia*, 1 June 1779, leg. 1261, PC; "Noticia de los Vecinos." The main names and other data on the 1770 *tazmias* match, with the exception of one entry; recorded in a span of less than three months by the same person, the later *tazmia* appears to be a revision. The production figures used here come from the later document.
12. Nicolás Hidalgo Gato, *tazmia*, 27 March 1771, San Juan y Martínez, leg. 1096, PC. Although Diego de Mesa and Captain Juan de Mesa shared the same surname, a connection between them is unknown at this time.
13. Ramón Lloret, *tazmia*, 1 June 1779, San Juan y Martínez, leg. 1261, PC; unsigned, *tazmia*, 20 June 1781, Filipinas, leg. 1317, PC; Gabriel de Mier, *tazmia*, 15 June 1782, Filipinas, leg. 1317, PC; Mier, *tazmia*, 11 June 1783, Filipinas, leg. 1350, PC.
14. María Asuncíon de los Reyes, testament, 30 January 1777, San Juan y Martínez, folio 97, exp. 71, book 1 (1775–1789), Protocolos Notariales (hereafter PN), Escribania de Gobierno, Archivo Historico Provincial de Pinar del Río, Cuba.
15. María Asuncíon de los Reyes, testament, 19 October 1785, folio 484, exp. 309, book 1 (1775–1789), PN. De los Reyes specified that the animals should be left to the local *capitán del partido*, Nicolás Hidalgo Gato.
16. De Lamadriz, "Resumen general"; José Cipriano de Ignocencio to Domingo González, sale of slave, 29 May 1775, San Juan y Martínez, exp. 7, book 1 (1776–1789), PN; Gabriel de Mier, *tazmia*, 15 June 1782, Filipinas, leg. 1317, PC; Bernardo Romero, *tazmia*, 22 April 1782, Consolación del Sur, leg. 1261, PC; Ramón Lloret, *tazmia*, 24 April 1780, Filipinas, exp. 156, folio 228v, leg. 1261, PC; Romero, *tazmia*, 20 June 1781, Consolación del Sur, leg. 1317, PC; Mier, *tazmia*, 20 June 1781, Filipinas, leg. 1317, PC; Lloret, *tazmia*, 1 June 1779, San Juan y Martínez, leg. 1261, PC; Mier, *tazmia*, 11 June 1783, Filipinas, leg. 1350, PC.
17. Marrero, *Cuba*, 9:215.
18. "Memoria del reconocimiento de la parte occidental de esta Isla. . . ." *Memorias de la Real Sociedad Patriótica de La Habana* 2, no. 25: 430, quoted in Marrero, *Cuba*, 9:216.

19. De la Sagra, *Historia economico-política y estadistica*, 3, cited in Kenneth F. Kiple, *Blacks in Colonial Cuba, 1774–1899*, Latin American Monographs, 2nd Series (Gainesville: University Presses of Florida, 1976), "Appendix: Cuba's Official Censuses, 1774–1899," census of 1774, n.p.
20. Jacinto Morejon to María Loreto, letter of freedom, 18 July 1788, Consolación, exp. 413, book 1 (1775–1789), PN; Domingo Duarte, *tazmia*, Consolación, 1770, leg. 1094, PC; Ramón Lloret, *tazmia*, 1 June 1779, leg. 1261, PC.
21. José Antonio Ribero to don Anzieto Morejon, sale, 27 March 1786, Filipinas, folio 503v, exp. 317, book 1 (1775–1789), PN; María Loreto to Morejon, sale, 18 October 1788, Filipinas, folio 634v, book 1 (1775–1789), PN; Morejon to Loreto, letter of freedom; Morejon to Loreto, obligation, 18 October 1788, Pinar del Río, folio 636v, exp. 423, book 1 (1775–1789), PN. For Bartolomé Diaz as a *veguero* and slave owner, see José de la Torre to Sgt. Felipe Rocha, sale, 17 September 1778, folio 190v, exp. 119, book 1 (1775–1789), PN; and Antonio de Puentes to Bartolomé Diaz, sale, 24 November 1780, exp. 149, book 1 (1775–1789), PN.
22. Gwendolyn Midlo Hall notes that Africans as well as Europeans identified their slaves by ethnic designations and descriptions. Mina, she notes, were not those Africans who were transported via the Gold Coast's São Jorge da Mina (Elmina) slave castle, but rather "a designation referring to different ethnicities over time and place" (*Slavery and African Ethnicities in the Americas: Restoring the Links* [Chapel Hill: University of North Carolina Press, 2005], 47).
23. Domingo González, declaration, 24 April 1775, San Juan y Martínez, exp. 3, book 1 (1775–1789), PN; González to Antonio Bentura, letter of freedom, 17 October 1775, San Juan y Martínez, exp. 27, book 1 (1775–1789), PN.
24. Kiple, *Blacks in Colonial Cuba*, appendix. For slave labor, manumissions, and the role of gender, see Marrero, *Cuba*, 9:178–210; Knight, *Slave Society in Cuba*, 78–79; Mary Karasch, "Slave Women on the Brazilian Frontier in the Nineteenth Century," in *More Than Chattel: Black Women and Slavery in the Americas*, edited by David Barry Gaspar and Darlene Clark Hine (Bloomington: Indiana University Press, 1996), 87–88; Frederick P. Bowser, "The Free Person of Color in Mexico City and Lima: Manumission and Opportunity, 1580–1650," in *Race and Slavery in the Western Hemisphere: Quantitative Studies*, edited by Stanley L. Engerman and Eugene D. Genovese (Princeton. NJ: Princeton University Press, 1975), 331–68.
25. Manuel Hernández González, *Canarias: La emigración* (La Laguna, Tenerife: Impecan, 1995), 20; James J. Parsons, *Hispanic Lands and Peoples: Selected Writings of James J. Parsons*, edited by William M. Denevan, Dellplain Latin American Studies (Boulder, CO: Westview Press, 1989), 400, 403; Julio Le Riverend Brusone, *Historia económica de Cuba* (Barcelona: Ediciones Ariel, 1972), 56; John Robert McNeill, *Atlantic Empires of France and Spain: Louisbourg and Havana, 1700–1763* (Chapel Hill: University of North Carolina Press, 1985), 35; Julio Hernández García, *La emigración canaria contemporánea (Siglo XIX)* (Las Palmas, Spain: Ediciones del Cabildo Insular de Gran Canaria, 1987), 32–33; Antonio M. Macías Hernández, *La migración canaria, 1500–1980* (Colombres, Spain: Ediciones Júcar, 1992), 84–85, 124–25; Antonio Lorenzo, Marcial Morera, and Gonzalo Ortega, *Diccionario de canarismos* (La Laguna, Tenerife: Francisco Lemus, 1995), 343. Canary Islanders were not always welcomed, as demonstrated by *real cédulas* in 1574 and 1599, which prohibited *isleño* immigration to the Indies (Macías Hernández, *La migración canaria*, 41).

26. Manuel Ramos, *tazmia*, n.d., Canoa, leg. 1094, PC; Joaquín de Sola, *tazmia*, 25 May 1770, Puerto Príncipe, leg. 1085, PC; Vicente del Castillo, *tazmia*, 3 May 1772, Cano, leg. 1196, PC; Simón de la Cruz, *tazmia*, 7 May 1773, Govea, leg. 1196, PC; del Castillo, *tazmia*, 10 July 1774, Cano, leg. 1192, PC; de Sola, *tazmia*, 24 March 1777, Puerto Príncipe, leg. 1180, PC; Fernando Muñoz, *tazmia*, 7 July 1781, Trinidad, leg. 1313, PC; Gabriel de Mier, *tazmia*, 15 June 1782, Filipinas, leg. 1317, PC; Muñoz, *tazmia*, 2 April 1783, Trinidad, leg. 1349, PC. These figures are based on formulas found in José de Aguilar y Blanco, *tazmia*, 20 March 1782, Sancti Spíritus, leg. 1313, PC; Manuel del Castillo, "Relasion de las Vegas . . . Río de Cuiaguatexe," 30 May 1771, Cuiaguatexe, leg. 1096, PC; and "Autos sobre el mal uso que hacen los vegueros. . . ." 22 November 1798, San Juan de Remedios, no. 7, leg. 1003, IGH.
27. Joaquín de Sola, *tazmia*, 24 March 1777, Puerto Príncipe, leg. 1180, PC; Fernando Muñoz, *tazmia*, 7 June 1781, Trinidad, leg. 1313, PC; Muñoz, "Notisia . . . el numero de vegas," 2 April 1783, Trinidad, leg. 1349, PC; de Sola, *tazmia*, 25 May 1770, Puerto Príncipe, leg. 1095, PC; Manuel Ramos, *tazmia*, n.d., Canoa, leg. 1094, PC; Vicente del Castillo, *tazmia*, 10 July 1774, Cano, leg. 1192, PC; Nobiatur, *El veguero*, 15; Un Amigo del Pais, *Memoria sobre el tabaco*, 25; Antonio María de Paula Arias, *El veguero de Vuelta Abajo: Apuntes sobre el cultivo del tabaco* (Pinar del Río, Cuba: Imprenta de Miguel Vives, 1887), 22.
28. José Antonio González, testimony, 9 May 1776, Santiago de las Vegas, Escribania de Gobierno (hereafter EG), Carlos Ximénez 1776–1777, pages 43–48, ANC.
29. *Real Cédula de 21 de octubre de 1817, sobre aumentar la población blanca de la isla de Cuba* (Havana: Oficina de Arazoa y Soler, 1818), section 18, folder 5, Spanish Consulado (SC) at Norfolk, VHS; Susan Schroeder, *Cuba: A Handbook of Historical Statistics* (Boston: G. K. Hall, 1982), 106; Macías Hernández, *La migración canaria*, 96; Conde Armíldez de Toledo, *Noticias estadisticas de la Isla de Cuba en 1862, dispuestas y publicadas por el centro de estadistica* (Havana: Imprenta del Gobierno, Capitania General y Real Hacienda por S.M., 1864), 46. A later royal order offered even more enticing terms to foreign immigrants, including the possibility of free land. See *Real orden que espidieron las Córtes Ordinarias en 27 de junio de 1821, y sancionó S.M. en 12 de marzo de 1822, sobre el aumento de la población blanca de estangeros en las provincias españolas ultramarinas* (Havana: Imprenta Fraternal de los Diaz de Castro, 1822), section 18, folder 7, SC.
30. Francisco Nates to Marqués de la Torre, 24 April 1772, Havana, leg. 1192, PC. The exchange rate is calculated using Rafael Torres Sánchez, Javier Gómez Biscarri, and Fernando Pérez de Gracia, "Exchange Rate Behavior and Exchange Rate Puzzles: Why the 18th Century Might Help," *Revista de Historia Económica-Madrid* 23, no. 1 (2005): 154; Thomas E. Chávez, *Spain and the Independence of the United States: An Intrinsic Gift* (Albuquerque: University of New Mexico Press, 2002), "Appendix 3: Money at the Time of Independence of the United States," 224.
31. Felipe Bellos to Antonio María de Bucarely, 24 August 1766, Sancti Spíritus, doc. 10, leg. 1087, PC; Martín de Asanza to de la Torre, 7 July 1774, Bayamo, leg. 1179, PC; Juan Antonio Ayans de Ureta to Pascual de Cisneros and de la Torre, 4 September 1774, Santiago de Cuba, leg. 1142, PC; Alejandro Ramírez to Secretario de Estado y del Despachado de Hacienda de España, 10 October 1817, Havana, leg. 2019, SD; Julián de Arriaga to Bucarely, 22 September 1770, San Ildefonso, folio 119, leg. 1134, PC.

32. Pedro Suarez de Urbina to Secretario de Estado y Despacho de Hacienda, 14 December 1813, Santiago de Cuba, leg. 2001, SD; Suarez de Urbina to Intendente del Ramo de Tabaco, 18 October 1813, Santiago de Cuba, leg. 2001, SD.
33. Marqués de la Torre to Martín de Asanza, 4 July 1774, Havana, leg. 1179, PC. De la Torre's letter reads: "Me hago cargo de la necesidad que obliga a UM a solicitar el permiso de buscar en ese vencindario algunas cantidades de dinero, librandolas contra la Tesoreria del Ramo de Tabaco de esta ciudad, para contentan en alguna manera a los vegueros de esa Juirsdiccion que solo con el producto de sus cosechas pueden aliviar las miserias que padacen; y sobre este particular se arreglara UM a lo que le perviniere el Administrador General de la Factoría."
34. Rafael Gómez Robaud to Francisco de Saavedra, 9 January 1810, Havana, leg. 2001, SD. Officials as early as the 1760s had noted that prompt payment would help produce a good crop. Julian de Arriaga to Antonio María Bucarely, San Ildefonso, 22 August 1769, folio 50, leg. 1134, PC.
35. Tomás de la Luz to Antonio María de Bucarely, 25 May 1768, Pinar del Río, leg. 1093, PC; Antonio Fernández to Marqués de la Torre, 13 March 1775, Filipinas, leg. 1180, PC; Fernández to de la Torre, 29 April 1775, Filipinas, leg. 1180, PC.
36. Nicolás Ramos et al. to Teniente Gobernador, [1780], Filipinas, leg. 1261, PC.
37. Ramón Lloret to Diego José Navarro, 31 May 1780, Filipinas, leg. 1261, PC. The original reads as follows: "Que tengo que hazer con un Hombre Veguero, con familia que despues de haver corrido todos los Hatos, y Corrales de esta Jurisdicíon, a fín de que le vendan una res, o un serdo, llevando su dínero en la mano (y esto Sor porque Yo de mí bosillo, por conocer su nesecidad, les hago la caridad de prestarselo) no encuentra quíen le Venda una Libra de Carne."
38. Diego José Navarro to Gabriel de Mier, 26 August 1780, Havana, leg. 1261, PC; Mier to Navarro, 9 May 1781, Filipinas, leg. 1317, PC; Navarro to Mier, 11 June 1781, Havana, leg. 1317, PC; "Todos los oficios de D. José María de la Torre," report and attachments, 12 March 1805, leg. 2018, SD.
39. Hubert H. S. Aimes, *A History of Slavery in Cuba, 1511–1868* (New York: Octagon Books, 1967), 23–24, 27.
40. Alonso Arcos y Moreno to Francisco Cagigal de la Vega, 13 December 1752, Santiago, leg. 6, CG.
41. Francisco Cagigal de la Vega to Alonso Arcos y Moreno, 10 March 1753, Havana, leg. 6, CG.
42. Luis de Echavarría to Conde de Ricla, 16 April 1764, Santiago de Cuba, order 5, leg. 6, CG.
43. Julián de Arriaga to Antonio María de Bucarely, 22 September 1769, San Ildefonso, folio 46, leg. 1134, PC; Miguel de Múzquiz to Marqués de la Torre, 7 July 1773, Madrid, leg. 1227, PC; Martín de Asanza to de la Torre, 15 August 1774, Bayamo, exp. 69, leg. 1174, PC; de la Torre to Asanza, 23 September 1774, Bayamo, leg. 1179, PC. Cuba's and Spain's needs in the early 1770s were at odds at this particular time. Seville had a surplus of tobacco, and neither an increase in the number of tobacco farms or production was desired by the metropole.
44. Aimes, *Slavery in Cuba*, 26–27, 30; Sherry Johnson, "Where Has All the Flour Gone? El Niño, Environmental Crisis and the Emergence of Alternative Markets in the Hispanic Caribbean, 1760s–1770s," *William and Mary Quarterly* 62, no. 3 (July 2005): 365–410; Bibiano Torres Ramírez, *La Compañía Gaditana de Negros*

(Seville: Escuela de los Estudios Hispano-Americanos de Sevilla, 1973). Villanueva Pico's words are from Aimes, *Slavery in Cuba*, 30.

45. "Nota de las cavesas de negros. . . ." 21 March 1764, leg. 1212, SD; Martín Xavier de Echavarría to Marqués de la Torre, 19 May 1772, Havana, leg. 1156, PC; unknown to Miguel de Múzquiz, 9 October 1773, Havana, leg. 1227, PC; de la Torre to Martín de Asanza, 23 September 1774, Havana, leg. 1179, PC; Francisco Jacott, "Cuenta de cargo y data de los caudales de la Real Factoría. . . ." 26 July 1804, Havana, leg. 250, Ultramar, AGI. The slaves sold in 1764 for approximately fifty *pesos* each—less than half the normal price.
46. Letter and testimony, 9 September 1801, no. 190, leg. 2000, SD; letter and testimony, 25 June 1802, leg. 2000, SD; Luis de Viguri to Miguel Cayetano Soler, 28 June 1800, Havana, leg. 2000, SD; "Junta Acuerdo," 25 January 1800, leg. 2000, SD; Aimes, *Slavery in Cuba*, 23–24, 27.
47. Martín Xavier de Echavarría to de la Torre, 19 May 1772, Havana, leg. 1156, PC; "Expediente de oficio contra don Agustín Ruiz. . . ." 21 July 1792, no. 9, leg. 1003, IGH. There is a discrepancy about the year that Ruiz purchased the slaves. The 1772 document states that the sale took place in 1769, while the 1792 document claims that it took place in 1765.
48. "Partido de Güines, Estado General. . . ." 22 August 1774, Güines, photocopy in *Población*; Knight, *Slave Society in Cuba*, 150–51. In both Cuba and the US South, slaves lived and worked alongside a limited number of their enslaved brethren. Before the US Civil War, most Southern slaves—approximately 75 percent—lived in groups of between one and forty-nine. See Peter Kolchin, *American Slavery, 1619–1877* (New York: Hill and Wang, 1993), 101.
49. The description of a slave's average workday on the Vuelta Abajo's *vegas* is derived from Miguel Rodríguez-Ferrer, *El tabaco habano: Su historia, su cultivo, sus vicisitudes, sus más afamadas vegas en Cuba* (Madrid: Imprenta del Colegio Nacional de Sordo-Mudos, 1851), 95–97; part of Rodríguez-Ferrer's description is quoted in Marrero, *Cuba*, 11:63.
50. Marrero, *Cuba*, 11:63.
51. Mateo de Echavarría y Elguerra, "Relación de los Yndividuos Labradores. . . ." 13 September 1768, Santiago de Cuba, exp. 41, leg. 23, CG; José E. Perdomo, *Léxico tabacalero cubano* (Miami: Ediciones Universal, 1998 [1940]), 191, 204; Fernando Muñoz, *tazmia*, 3 April 1773, Trinidad, exp. 44, leg. 1186, PC; Muñoz, *tazmia*, 13 April 1774, Trinidad, exp. 57, leg. 1186, PC; Muñoz, *tazmia*, 9 June 1775, Trinidad, exp. 67, leg. 1186, PC; unsigned, *tazmia*, 24 May 1776, Trinidad, exp. 74, leg. 1186, PC.
52. Vicente del Castillo, *tazmia*, 5 May 1771, Cano, leg. 1096, PC; del Castillo, *tazmia*, 3 May 1772, Cano, leg. 1196, PC. The 1772 slave owners were Antonio de Lima, José Pérez, don Matias Cantos, and Felipe de Torres.
53. Perdomo, *Léxico tabacalero cubano*, 184. For more on slave *conucos* throughout the Americas, see Herbert S. Klein, *Slavery in the Americas: A Comparative Study of Virginia and Cuba* (Chicago: University of Chicago Press, 1967), 99; Stuart B. Schwartz, *Slaves, Peasants, and Rebels: Reconsidering Brazilian Slavery* (Urbana: University of Illinois Press, 1992), 45–48; B. J. Barickman, "A Bit of Land, Which They Call *Roca*: Slave Provision Grounds in the Bahian Reconcavo, 1780–1860," *Hispanic American Historical Review* 74, no. 4 (November 1994): 649–87; and Woodville K. Marshall,

"Provision-Ground and Plantation Labour in Four Windward Islands: Competition for Resources During Slavery," *Slavery Abolition* 12, no. 1 (May 1991): 48–67. Klein notes that Cuban slaves were allowed to work their *conucos* on Sundays and holy days. The income from these garden plots were exempt from tithe payments, "an extremely unusual privilege in colonial society, where the *diezmos*, or tithes, were the most universal form of production and property taxes" (*Slavery in the Americas*, 99).

54. Nicolás Hidalgo Gato, "Tazmia . . . Partido de San Juan y Martínez . . . Partido del Sur," 22 June 1770, San Juan y Martínez, leg. 1094, PC; Manuel del Castillo, "Relasion de la Vegas"; Ramón Lloret, "Tasmia . . . algunos Partidos de la Buelta de abajo. . . ." 24 April 1780, Filipinas, exp. 156, folio 228v, leg. 1261, PC; Rebecca J. Scott, *Slave Emancipation in Cuba*, 263.

55. Ramón José Rodríguez, *tazmia*, 15 May 1770, Matanzas, leg. 1090, PC; Rafael Nieto y Cortadellas, *Dignidades nobiliarias en Cuba* (Madrid: Ediciones Cultura Hispanica, 1954), 250–53, 294–95. The tradition of using the former owners' surnames is well documented, particularly among slaves in the United States. See, for example, Herbert G. Gutman, *The Black Family in Slavery and Freedom, 1750–1925* (New York: Vintage, 1976).

56. Fernando Muñoz, *tazmia*, 3 April 1772, Trinidad, leg. 1186, PC; Muñoz, *tazmia*, 24 May 1776, Trinidad, exp. 74, leg. 1186, PC.

57. Allan J. Kuethe, *Cuba, 1753–1815: Crown, Military, and Society* (Knoxville: University of Tennessee Press, 1986), 38–41; Sherry Johnson, *Social Transformation*, 60; Juan Marchena Fernández, *Oficiales y soldados en el ejército de América* (Seville: Escuela de Estudios Hispanoamericanos, CSIC, 1983), 82. Kuethe notes that certain professions were exempt from the militia, including "those engaged in medical services, notaries, tax administrators, clerics, teachers, plantation foremen, and tobacco *factors*" (41).

58. José Rivero Muñiz, *Las tres sediciones de los vegueros en el siglo XVIII* (Habana: Academia de la Historia de Cuba, 1951), 27. The early protests against the monopoly are discussed in detail in Chapter 4. The presence of *vegueros* among militia ranks was not limited to just Cuba, as most of the tobacco farmers in South America's Río de la Plata region also served in the militia. See Francisco de Paula Sanz, *Viaje por el Virreinato del Río de la Plata: El camino del tabaco* (Buenos Aires: Centro de Estudios Interdisciplinarios de Hispanoamérica Colonial, 1977), 33–34.

59. Juan Miguel Arozena to Antonio María de Bucarely, 28 June 1766, Trinidad, exp. 8, leg. 1078, PC; Antonio Abad Brito to Ministro Factor, 21 April 1811, Holguín, order 423, leg. 5, CG; Juan de Mecolaeta and Pedro Antonio de Gamon to Capitán General, 26 August 1799, leg. 2018, SD; Juan Francisco Gion to Juan de Mecolaeta, 15 March 1801, Mayarí, leg. 235, Ultramar, AGI; Miguel Villalon y Rivera to Mecolaeta, 15 April 1801, Santiago de Cuba, leg. 235, Ultramar, AGI; Marqués de Guisa to Mecolaeta, 14 May 1801, Bayamo, leg. 235, Ultramar, AGI.

60. Allan J. Kuethe, "The Development of the Cuban Military as a Sociopolitical Elite, 1763–83," *Hispanic American Historical Review* 61, no. 4 (1981): 700; Kuethe, *Cuba*, 44–45.

61. Sherry Johnson, *Social Transformation*, 64–65; Marqués de la Torre to Gaspar Gelabert, 30 June 1772, Havana, exp. 24, leg. 1186, PC.

62. Lorenzo de Madriaga to Juan Miguel Portuando, order, 2 January 1755, Santiago

de Cuba, order 270, leg. 7, CG; Luis de Echavarría, "Lista de los milicianos. . . ." 9 July 1765, Santiago de Cuba, order 10, leg. 26, CG; Luis de Echavarría, *tazmia*, 12 April 1770, Mayarí, leg. 1080, PC.

63. Matias Joseph Duarte, *tazmia*, 6 June 1770, San Ildefonso de Guane, leg. 1094, PC; Manuel del Castillo, "Relasion de la Vegas"; Mateo Morgado to Antonio Sánchez, sale, 19 February 1780, Filipinas, folio 228v, exp. 156, book 1 (1775–1789), PN; Luis Dias Pimienta, "Noticia de los Vezinos que comprehende el partido de Guane del Sur y como capitán doy al Señor Gobernador y Capitán General con arreglo a su orden de tienta y uno de Marzo de mil setecientos ochenta y seis," 20 April 1786, Partido de Guane del Sur, photocopy in *Población*.

64. Ana Sánchez to Mateo Morgado, sale, 2 January 1777, Filipinas, exp. 73, book 1 (1775–1789), PN; Morgado to Sánchez, sale, 19 February 1780; Morgado to Rafael Ilobera, sale, 27 November 1781, Filipinas, folio 272v, exp. 185, book 2, PN.

65. Tomás de Sierra to Mateo Morgado, sale, 20 August 1793, Filipinas, folio 61, exp. 83, book 2, PN; Dias Pimienta, "Noticia de los Vezinos."

66. Sherry Johnson, *Social Transformation*, 39–70; Nicolás Ramos et al. to Teniente Gobernador, [1780], Filipinas, leg. 1261, PC; Dias Pimienta, "Noticia de los Vezinos"; Matias Joseph Duarte, *tazmia*, 6 June 1770, San Ildefonso de Guanes, leg. 1094, PC; Manuel del Castillo, "Relasion de las Vegas"; Gabriel de Mier, "Tasmia . . . algunos Partidos de la Buelta de Abajo. . . ." 20 June 1781, Filipinas, leg. 1317, PC; José Ramos to Teniente Bonafacio Guzman, *venta de asenso* (sale of consent), 22 July 1775, Filipinas, exp. 15, folio 13, book 1 (1775–1789), PN.

67. Bernabe Hernández, *tazmia*, 9 May 1770, Xiaraco, leg. 1094, PC; Hernández, *tazmia*, 1 June 1772, Xiaraco, leg. 1195, PC; Hernández, *tazmia*, 17 June 1773, Xiaraco, leg. 1195, PC; Hernández, *tazmia*, 16 June 1774, Xiaraco, leg. 1195, PC; Hernández, *tazmia*, 6 July 1775, Xiaraco, leg. 1195, PC; Hernández, *tazmia*, 7 July 1776, leg. 1193, PC; Conde de Casa Bayona to Luis de las Casas, 6 October 1791, Havana, leg. 1461, PC; Sherry Johnson, *Social Transformation*, 50.

68. Francisco Gama, census, 6 June 1767, Güines, leg. 1096, PC; "Noticio de los Vecinos . . . Guines. . . ." photocopy in *Población*; Gama, *tazmia*, 14 April 1768, Güines, leg. 1090, PC; Gama, *tazmia*, 15 May 1771, Güines, leg. 1090, PC; Gama, *tazmia*, 8 June 1774, Güines, leg. 1195, PC; Gama, *tazmia*, 16 June 1776, Güines, leg. 1195, PC. Del Pino may have temporarily stopped growing tobacco on one farm while continuing to do so on the other, allowing the land to lie fallow, a practice used in Virginia and Mexico but typically not in Cuba. See Meining, *Atlantic America, 1492–1800*, 154; Silver, *New Face*, 163; Brennan, *Tobacco Leaves*, 43; and Deans-Smith, *Bureaucrats, Planters, and Workers*, 116.

69. Much of the literature on the Catholic Church's agricultural pursuits in the New World has focused on the Jesuits and their activities. See Nicholas P. Cushner, *Lords of the Land: Sugar, Wine, and Jesuit Estates of Coastal Peru, 1600–1767* (Albany: State University of New York Press, 1980); Kendall W. Brown, "Jesuit Wealth and Economic Activity with the Peruvian Economy: The Case of Colonial Southern Peru," *Americas* 44, no. 1 (July 1987): 23–43; and Mercedes García Rodríguez, "Presencia Jesuita en la economía de Cuba: Siglo XVIII" (PhD diss., Universidad de la Habana, 1999), among others.

70. License for Dionisio Manrique, 13 November 1764, Madrid, folio 1, leg. 5508, Contratación; Manuel Pérez-Beato, *Toponimia*, vol. 1 of *Habana antigua* (Havana:

Seoane, Fernández, 1936), 202–3; Simón de la Cruz, *tazmia*, 21 June 1774, Govea, leg. 1196, PC. Under normal circumstances in colonial Cuba, 3,500 *cujes* was a large amount, but Batabanó must have been experiencing a particularly good year, as twenty-three *vegueros* out of the fifty-three in the entire district each produced at least one thousand *cujes* of tobacco.

71. Antonio Laso de la Vega, *tazmia*, 28 February 1773, Bayamo, leg. 1174, AGI. The estimate of two acres was based on calculations as to the number of tobacco plants that could be grown in a given area. Several estimates indicated that an average of five thousand plants could be grown on an acre. See Susan Deans-Smith, "The Money Plant: The Royal Tobacco Monopoly of New Spain, 1765–1821," in *The Economies of Mexico and Peru during the Late Colonial Period, 1760–1810*, edited by Nils Jacobsen and Hans-Jürgen Puhle (Berlin: Colloquium-Verlag, 1986), 386.

72. Moreno Fraginals, *Sugarmill*, 21; Marrero, *Cuba*, 11:22–23. The junior Núñez del Castillo's offer of land to be used for tobacco is discussed in greater detail later in this chapter. Those with noble titles and ties to tobacco were the Marqués de Casa Torres, Laureano de Torres Ayala; the aforementioned elder Marqués de San Felipe y Santiago; the Conde de Casa Bayona, Captain José Bayona y Chacón; the Marqués de Jústiz y Santa Ana, Juan José de Jústiz y Umpirre; and the Marqués del Real Agrado, Domingo de Lizundia y Odria. See Leví Marrero, *Azúcar, ilustración y conciencia (1763–1868) (V)*, vol. 13 of *Cuba: Economía y sociedad* (Madrid: Editorial Playor, 1986), 39–40.

73. Unsigned, *tazmia*, 10 July 1774, Partido del Cano, leg. 1192, PC; Marrero, *Cuba*, 13:39; Moreno Fraginals, *Sugarmill*, 21–22. Titles of Castile were noble titles issued by the Spanish Crown, frequently in return for service to Spain. See John Pinkerton, *A General Collection of the Best and Most Interesting Voyages and Travels, in All Parts of the World; Many of which Are Now Translated into English* (Philadelphia: Kimber and Conrad, 1812), 5:346, for a thorough description of the rights and privileges that came with this distinction.

74. Nieto y Cortadellas, *Dignidades nobiliarias*, 114–15; Leví Marrero, *Del monopolio hacia la libertad comercial (1701–1763) (III)*, vol. 8 of *Cuba: Economía y sociedad* (Madrid: Editorial Playor, 1980), 167; Kuethe, *Cuba*, 52, 57.

75. Kuethe, *Cuba*, 57; Bernabé Hernández, *tazmia*, 6 July 1775, Xiaraco, leg. 1195, PC; Hernández, *tazmia*, 1 June 1772, Xiaraco, leg. 1195, PC; Hernández, *tazmia*, 17 June 1773, Xiaraco, leg. 1195, PC; Hernández, *tazmia*, 16 June 1774, Xiaraco, leg. 1195, PC; Hernández, *tazmia*, 7 July 1776, Xiaraco, leg. 1193, PC.

76. Bernabé Hernández, *tazmia*, 1 June 1772, Xiaraco, leg. 1195, PC; Hernández, *tazmia*, 17 June 1773, Xiaraco, leg. 1195, PC; Hernández, *tazmia*, 16 June 1774, leg. 1195, PC; Hernández, *tazmia*, 7 July 1776, Xiaraco, leg. 1193, PC; Manuel Ramírez de Arellano, "Estado que manifiesta los Tabacos que han entragado los Labradores. . . ." 4 August 1804, Real Factoría de Tabacos, Havana, table 9, signature 9454, leg. 209, Real Consulado y Junta de Fomento (hereafter RC), ANC; Ramirez de Arellano, "Estado que manifiesta los distintos precios. . . ." 4 August 1804, Real Factoría de Tabacos de la Havana, table 10, signature 9454, leg. 209, RC, and leg. 2001, SD; Sherry Johnson, *Climate and Catastrophe*.

77. Miguel de Múzquiz to Marqués de la Torre, 25 December 1775, Madrid, doc. 52,

folios 884–86, leg. 1227, PC; Julián de Arriaga to Antonio María de Bucarely, 18 July 1769, Madrid, leg. 1134, PC; Arriaga to Bucarely, 22 October 1769, San Lorenzo, folios 54–56, leg. 1134, PC; Arriaga to Bucarely, 22 June 1770, Aranjuez, folio 100, leg. 1134, PC; Juan Baptista Bonet to de la Torre, 4 February 1775, leg. 1178, PC.

78. Rafael Gómez Roubaud to Miguel Cayetano Soler, 17 February 1806, Havana, leg. 2000, SD; "Repartimiento de las cien cavallerías de tierra. . . ." leg. 670, Mapas y Planos (hereafter MP), SD; Junta, 3 January 1800, leg. 2018, SD.

79. Rafael Gómez Roubaud to Miguel Cayetano Soler, 31 July 1806, Havana, leg. 198, Ultramar, AGI.

80. Rafael Gómez Roubaud to Miguel Cayetano Soler, 17 April 1805, Havana, leg. 253, Ultramar, AGI.

81. Rafael Gómez Roubaud to Príncipe de la Paz (Manuel Godoy), 29 March 1807, Havana, leg. 198, Ultramar, AGI.

82. Rafael Gómez Roubaud to Miguel Cayetano Soler, 12 April 1807, Havana, leg. 198, Ultramar, AGI; Príncipe de la Paz to Soler, Madrid, 20 June 1807, leg. 198, Ultramar, AGI; Kuethe, *Cuba*, 152–54. For more on the count's titles and connections, see Chapter 5.

83. Juan Tio, "Noticia del tavaco en Rama de varios Partidos que sea reconizido, segun e manifestado en esta Rl Fuerza a saver," 13 May 1760, Havana, leg. 2016, SD. For societal regulations for women, see Steve J. Stern, *The Secret History of Gender: Women, Men, and Power in Late Colonial Mexico* (Chapel Hill: University of North Carolina Press, 1995); Ann Twinam, *Public Lives, Private Secrets: Gender, Honor, Sexuality, and Illegitimacy in Colonial Spanish America* (Stanford, CA: Stanford University Press, 1999); Verena Martínez-Alier, *Marriage, Class and Colour in Nineteenth-Century Cuba* (Cambridge: Cambridge University Press, 1974); and Luis Martínez-Fernández, "The 'Male City' of Havana: The Coexisting Logics of Colonialism, Slavery, and Patriarchy in Nineteenth-Century Cuba," in *Women and the Colonial Gaze*, edited by Tamara L. Hunt and Micheline R. Lessard (New York: New York University Press, 2002), 104–16.

84. Sherry Johnson, "'*Señoras en Sus Clases no Ordinarias*': Enemy Collaborators or Courageous Defenders of the Family?" *Cuban Studies/Estudios Cubanos* 34 (2004): 25–30; Agustín Hernández Pulgarón, "Estado . . . Partidos de Bacuranado, San Jeronimo y Baxunallado. . . ." 6 April 1770, Guanabacoa, leg. 1090, PC; Simón de la Cruz, *tazmia*, 21 June 1774, Govea, leg. 1196, PC; Ramón José Rodríguez, *tazmia*, 26 May 1770, Matanzas, leg. 1090, PC; Rodríguez, *tazmia*, 12 May 1771, Matanzas, leg. 1090, PC.

85. Deans-Smith, *Bureaucrats, Planters, and Workers*, 11, 87, 111, 386.

CHAPTER 4

1. For tobacco monopolies in other locations, see, for example, Norton, *Sacred Gifts*, 209–28; Todd Butler, "Power in Smoke: The Language of Tobacco and Authority in Caroline England," *Studies in Philology* 106, no. 1 (Winter 2009), 110–18; William Brantley Aycock, "Tobacco Regulation in Colonial Virginia" (MA thesis, University of North Carolina, 1937).
2. Aycock, "Tobacco Regulation," 12, 17–35; Goodman, *Tobacco in History*, 76.

3. Much of my thinking on the paternal link between the Bourbon monarchy, its colonies, and the Bourbon reforms was shaped by Steinar A. Saether, "Bourbon Absolutism and Marriage Reform in Late Colonial America," *Americas* 59, no. 4 (April 2003): 475–509. Laura Náter convincingly argues that the initial Castilian tobacco monopoly was established in response to a dire need for revenue, prompted by European wars that began in the early part of the seventeenth century ("Colonial Tobacco," 103).
4. Contraband and other forms of subtle resistance are described in detail in a later chapter. The definitive work on subtle forms of resistance is James C. Scott, *Domination and the Arts of Resistance: Hidden Transcripts* (New Haven, CT: Yale University Press, 1990).
5. Náter, "Colonial Tobacco," 96, 98–99, 103. Measures to control tobacco were extended to England's American colonies as early as 1619, when Virginia passed statutes to ensure that, despite the popularity of tobacco cultivation, the colony maintained some sort of agricultural diversity. By comparison, Spain's tobacco monopoly was relatively late (Aycock, "Tobacco Regulation," 12, 17–35; Náter, "Colonial Tobacco," 96).
6. Diego Suarez de Amaya, quoted in Willem Wubbo Klooster, "Illicit Riches: The Dutch Trade in the Caribbean, 1648–1795" (PhD diss., Rijksuniversiteit te Leiden, Netherlands, 1995), 26.
7. Marcy Norton, "The Business of Tobacco in the Spanish Empire, 1590–1636" (Working Paper 99-00, International Seminar on the History of the Atlantic World, 1500–1800, Harvard University, Cambridge, MA, 1998), 14–15.
8. David Lorne McWaters, "The Royal Tobacco Monopoly in Bourbon Mexico, 1764–1810" (PhD diss., University of Florida, 1979), 11–12, 24, 30; Ed. C. de Jesus, *The Tobacco Monopoly in the Philippines: Bureaucratic Enterprise and Social Change, 1760–1880* (Manila: Ateneo de Manila University Press, 1980), 29. The succession of the Bourbons to the Spanish throne was a complicated one. See John D. Bergamini, *The Spanish Bourbons: The History of a Tenacious Dynasty* (New York: Putnam, 1974), 21–37, for the difficulties involved in naming a new heir. Throughout much of the colonial period, Mexico was responsible for providing funds to support other parts of the Spanish Empire, including Cuba. Although the amount varied over time, these funds were necessary to pay for key components of Spain's occupation and control of other areas of the empire, such as fortifications, support of the garrisons, and in this case, money to pay tobacco growers. See, for example, Allan J. Kuethe, "El situado mexicano, los azucareros y la fielded cubana," in *Las Antillas en la era de las luces y la revolución*, edited by José Antonio Piqueras Arenas (Madrid: Siglo XXI de España Editores, 2005), 301–18; Altagracia Ortiz, *Eighteenth-Century Reforms in the Caribbean: Miguel de Muesas, Governor of Puerto Rico, 1769–76*, (East Brunswick, NJ: Associated University Presses, 1983), 54–55.
9. Norton, *Sacred Gifts*, 155, 162–63, 201; Goodman, *Tobacco in History* 146; McNeill, *Atlantic Empires*, 115.
10. McNeill, *Atlantic Empires*, 118; Marrero, *Cuba*, 7:45–46; Felipe V, "Instrucción dada por Felipe V el día 11 de abril de 1717 para el estanco de tabaco," in *Documentos para la historia de Cuba*, vol. 1, edited by Hortensia Pichardo (Havana: Editorial de Ciencias Sociales, 1971), 147; Rivero Muñiz, *Las tres sediciones*, 20–22.
11. Rivero Muñiz, *Tabaco*, 1:90–94; McNeill, *Atlantic Empires*, 118, 256. McNeill argues,

however, that the fixed prices were "notably lower" (256) than those paid to Cuban tobacco farmers in 1700.

12. Rivero Muñiz, *Las tres sediciones*, 27; Guerra y Sánchez et al., *Historia*, 2:263; Asiris M. Cruz, "Smoking Guns: Tobacco Farmers' Revolts and Colonial Society, Cuba, 1650–1750" (PhD diss., University of California, Riverside, 2007), 102.
13. Rivero Muñiz, *Tabaco*, 1:90, 1:105–11; Rivero Muñiz, *Las tres sediciones*, 27, 33–39; Cruz, "Smoking Guns," 106–7.
14. Cruz, "Smoking Guns," 107–9; Marrero, *Cuba*, 7:50; Gregorio Guazo Calderón, license, no. 1, leg. 5470, Contratación.
15. Rivero Muñiz, *Las tres sediciones*, 59–63.
16. Ibid; Cruz, "Smoking Guns," 177–79.
17. Rivero Muñiz, *Las tres sediciones*, 67, 69–71.
18. Ibid.
19. Rivero Muñiz, *Tabaco*, 1:119, 1:122–24; Rivero Muñiz, *Las tres sediciones*, 71–75, 81–93; Marrero, *Cuba*, 7:45–54; Cruz, "Smoking Guns," 278–80.
20. Cruz, "Smoking Guns," 280.
21. Rivero Muñiz, *Tabaco*, 1:119, 122–24; Rivero Muñiz, *Las tres sediciones*, 91–93; Cruz, "Smoking Guns," 278–80.
22. Charles M. Andrews, ed., *Narratives of the Insurrections, 1675–1690* (New York: Barnes and Noble, 1967 [1915]), 11; Harry Firestone, ed., *Bacon's Rebellion: The Contemporary News Sheets* (Charlottesville: University of Virginia Press, 1956), 29; Aycock, "Tobacco Regulation," 71–82; K. G. Davies, *The North Atlantic World in the Seventeenth Century* (Minneapolis: University of Minnesota Press, 1974), 226; Theodore Saloutos, "Efforts at Crop Control in Seventeenth Century America," *Journal of Southern History* 12, no. 1 (February 1946): 62–65. Aycock contends that "to interpret the causes which led to Bacon's Rebellion in terms of tobacco would not be varying a great deal from the actual situation" ("Tobacco Regulation," 72). Andrews describes Bacon's Rebellion as "the first popular uprising in colonial America" (*Narratives*, 11).
23. Kulikoff, *Tobacco and Slaves*, 109–12.
24. Marcello Carmagnani, "La oposición a los tributos en la segunda mitad del siglo XVIII," *Revista chilena de historia y geografia* 129 (1961): 161–68; John P. Harrison, "The Evolution of the Colombian Tobacco Trade, to 1875," *Hispanic American Historical Review* 32, no. 2 (May 1952): 166. Carmagnani notes that the 1776 revolt was one of the first by Chilean creoles against Spain, while Harrison describes the *comunero* revolt as "the only spontaneous uprising of any size in the colonial history of New Granada." Those who participated in this rebellion described themselves as *comuneros* (commoners).
25. Marrero, *Cuba*, 7:52–56; McNeill, *Atlantic Empires*, 115; Jacobo de la Pezuela, "Tabaco," in *Diccionario geográfico, estadistico, histórico de la Isla de Cuba*, vol. 4 (Madrid: Imprenta del Establecimiento de Mallado, 1867), 566.
26. Marrero, *Cuba*, 7:58; McNeill, *Atlantic Empires*, 119–21; "Estableciendo una Compañía en San Cristóbal de la Habana Dedicada a la Conducción de Tabacos, Azúcar, Corambre y Otros Frutos de la Isla de Cuba," in *Cédulas de Luis I (1724), Cédulas de Felipe V (1724–1746)*, vol. 3 of *Cedulario Americano del Siglo XVIII*, edited by Antonio Muro Orejon (Seville: Escuela de Estudios Hispano-Americanos, 1740), 253–76; Rivero Muñiz, *Tabaco*, 1:225–36. Havana's Calle de Tallapiedra, a two-block-long

street located outside of the city's walls between Calle Suárez and Calle del Aguila, was named for tobacco monopoly head José Tallapiedra. See Pérez-Beato, *Toponimia*, 229.

27. "Articles of Capitulation," *Providence (RI) Gazette* 1, no. 3 (6 November 1762): 1; Rivero Muñiz, *Tabaco*, 1:237; Parcero Torre, *La pérdida*, 178, 183–84, 187.
28. Antonio Gutiérrez Escudero, "El tabaco en Santo Domingo y su exportación a Sevilla (época colonial)," in *Relaciones de poder y comercio colonial: Nuevas perspectivas*, edited by Enriqueta Vila Vilar and Allan J. Kuethe (Seville: Escuela de Estudios Hispano-Americanos de Sevilla, 1999), 124.
29. Marco Antonio Fallas, *La factoría de tabacos de Costa Rica* (San Jose: Editorial Costa Rica, 1972), 37, 61, 65; de Jesus, *Tobacco Monopoly*, 165.
30. Jacob M. Price, *France and the Chesapeake: A History of the French Tobacco Monopoly, 1674–1791, and of Its Relationship to the British and American Tobacco Trades* (Ann Arbor: University of Michigan Press, 1973), 303, 327–28, 357–59.
31. Brian E. Coutts, "Boom and Bust: The Rise and Fall of the Tobacco Industry in Spanish Louisiana, 1770–1790," *Americas* 42, no. 3 (January 1986): 289–91. New Spain during the colonial era encompassed much of what is now Central America and Mexico.
32. Ibid., 292.
33. Coutts, "Boom and Bust," 289–93, 295–309; Price, *France and the Chesapeake*, 79, 81.
34. Francisco Antonio de Astigarrota, "Informe sobre resguardo de a cavallo por la Administración principal de Renta Reales con comencuencia de auto de Señor Intendente Don José Pablo Valiente, Francisco Antonio de Astigarrota," 11 July 1787, Havana, no. 15, leg. 1034, IGH; Juan Ignacio de Urriza, "Nomina de los individuos que componen el Resguardo Rentas reales de la Ciudad de la Havana, y su campo, informe al Reglamento aprobado por S. M. con fecha de July de presente año," 28 September 1778, Havana, order 148, leg. 1114, IGH. A later document cites the original date of the *resguardo*'s founding as 1 October 1778. See José Garcia Alayeto, "Instrucciones para el resguardo montado de la Ciudad de la Havana creado el año de 1778 y reestablicido en este del 1787 en los terminos que en ella se refiere," 15 November 1787, Havana, no. 15, leg. 1034, IGH.
35. "Reglamento de los empleados. . . ." 30 August 1796, table 8, leg. 2001, SD; Rafael Gómez Roubaud, "Relación de todos los individuos. . . ." 8 April 1809, Havana, leg. 243, Ultramar, AGI; Joseph González, 31 May 1817, leg. 2019, SD.
36. Francisco del Nero, "Borrador de la Nomina de sueldos del mes de febrero de 1814," 26 February 1814, Seville, leg. 2123, AFTS. The *real* was also known as the *real de vellón*, and served as Spain's primary currency during the late eighteenth century. A *real* equaled about twenty US cents in 1792, according to Fernando Gutierrez, Carlos Larrinaga, and Miriam Núñez ("Cost and Management Accounting in Pre-industrial Revolution Spain," *Accounting Historians Journal* 32, no. 1 [June 2005], 148).
37. Jorge Luján Muñoz, "El establecimiento del Estanco," paper presented at the 5th Congreso Centroamericano de Historia (San Salvador, El Salvador, July 2000), 7–8; Eduardo Arcila Farias, *Historia de un monopolio: El estanco del tabaco en Venezuela 1779–1833* (Caracas: Ediciones de la Facultad de Humanidades y Educación, Universidad Central de Venezuela, 1977), 39–42; Deans-Smith, *Bureaucrats, Planters, and Workers*, 45–46.

38. Carlos Marichal and Matilde Souto Mantecón, "Silver and *Situados*: New Spain and the Financing of the Spanish Empire in the Caribbean in the Eighteenth Century," *Hispanic American Historical Review* 74, no. 4 (November 1994): 588, 601; Francisco Cagigal de la Vega to Alonso Arcos y Moreno, 30 April 1751, Havana, order 239, leg. 7, CG; Cagigal de la Vega to Arcos y Moreno, 30 April 1754, Havana, order 207, leg. 7, CG; Cagigal de la Vega to Arcos y Moreno, 30 April 1751, Havana, order 141, leg. 6, CG; Cagigal de la Vega to Lorenzo de Madariaga, 20 July 1755, Havana, order 89, leg. 7, CG; Madariaga to Cagigal de la Vega, 10 September 1755, Santiago de Cuba, order 255, leg. 7, CG; Arcos y Moreno to Cagigal de la Vega, 7 June 1752, Santiago de Cuba, order 234, leg. 6, CG. There were three types of silver coins circulating in the Spanish Empire during the late eighteenth century: double silver (*plata double*), and two types of fiduciary coins—the *de condorcillo* and the *macuquina*, both of which were also known as *plata sencilla*. Huberto M. Ennis, "Shortages of Small Change in Early Argentina," WP 03-12, Working Paper Series (Richmond, VA: Federal Reserve Bank of Richmond, 2003).
39. McNeill, *Atlantic Empires*, 158–61; John H. Coatsworth, "The Mexican Mining Industry in the Eighteenth Century," in Jacbosen and Puhle, *Economies*, 28.
40. Francisco Rodríguez Vizcayo, "Certifico que los dependientes que sirven en estas Rs Fábricas con sus sueldos anuales, graduaciones y destinos son los siguentes. . . ." 23 March 1762, Seville, leg. 2016, AFTS; Nero, "Borrador"; "Importe de 12 Nominas de Sueldos Correspondentes al presente año de 1817," n.d., leg. 2124, AFTS; Jacques A. Barbier and Herbert S. Klein, "Revolutionary Wars and Public Finances: The Madrid Treasury, 1784–1807," *Journal of Economic History*, 41, no. 2 (June 1981), 330.
41. McNeill, *Atlantic Empires*, 158; Manuel Ramírez de Arellano, "Estado que manifiesta el caudal que ha ingresado en Factoría. . . ." 4 August 1804, Real Factoría de Tabacos, Havana, table 11, signature 9454, leg. 209, RC; de Jesus, *Tobacco Monopoly*, 85.

CHAPTER 5

1. Marqués de la Torre to José Claro, 11 June 1773, Havana, leg. 1181, PC.
2. Thomas D. Boswell, "The West Indies: The Hispanic Territories and Haiti," in *Middle America: Its Lands and Peoples*, edited by Robert C. West et al. (Englewood Cliffs, NJ: Prentice Hall, 1989), 130, 134; R. H. Whitbeck, "Geographical Relations in the Development of Cuban Agriculture," *Geographical Review*, 12, no. 2 (April 1922), 236; *Sintesis de geografia de Cuba* (Miami: Ediciones del Directorio Magisterial Cubano [en el exilio], n.d.), 19, 22; Carlson, *Geography of Latin America*, 443, 445; Hazard, *Cuba*, 329.
3. [William Henry Hurlbert], *Gan-Eden; or, Pictures of Cuba* (Boston: John P. Jewett, 1854), 109. For other travelers who expressed similar sentiments about the Vuelta Abajo, see Abbot, *Letters*, 136; Maturin M. Ballou, *Due South; or, Cuba, Past and Present* (Boston: Houghton Mifflin, 1885), 248; Dimock, *Impressions*, 92; Hazard, *Cuba*, 329; Raymond A. Mohl, ed., "A Scotsman in Cuba, 1811–1812," *Americas* 29, no. 2 (1972): 245. Manuel Moreno Fraginals argues that the claim that the Vuelta Abajo had the world's best tobacco lands and tobacco was untrue. This, Moreno Fraginals asserts, was merely a method used by colonial officals to move tobacco cultivation west, away from prime sugar-growing areas around Havana (*El ingenio*, 60).

4. Santovenia, *Pinar del Río*, 53–54.
5. Marqués de la Torre to Julián de Arriaga, 22 January 1775, Havana, leg. 1220, PC, typescript copy in *Archivo General de Indias, AGI, DSC.*
6. Francisco Jacott to Marqués de la Torre, 4 September 1773, San Juan y Martínez, leg. 1220, PC, typescript copy in DSC. Jacott reported that the tobacco had been hidden in "uno en las barbacoas, otro en los aposentos, gallineros i chiqueros, i algunos en ranchos dentro del monte."
7. Francisco Jacott to Marqués de la Torre, 2 September 1773, San Juan y Martínez, leg. 1220, PC, typescript copy in DSC.
8. Francisco Jacott to Marqués de la Torre, 4 September 1773, San Juan y Martínez, leg. 1220, PC, typescript copy in DSC.
9. Marqués de la Torre to Julián de Arriaga, 22 January 1775, Havana, leg. 1220, PC, typescript copy in DSC; Santovenia, *Pinar del Río*, 55–57. De la Torre's letter to Arriaga reads, in part, "Cada día llegaban a mis oidos los clamores de los desordenes, injusticias y vexaciones de aquellos havitantes reciprocamente se inferian, y sufrian de los capitanes." Santovenia notes that the *partidos* were Consolación del Sur, Consolación del Norte, Pinar del Río, San Juan y Martínez, Guane del Sur, and Guane del Norte.
10. Tte. Col. Antonio Fernández, Capitán del Esquadrón de Dragones de America Fixo de Aquella Plaza, memorial, 19 November 1788, Havana, exp. 36, leg. 6881, Secretaría del Despacho de Guerra (hereafter SGU), Archivo General de Simancas (hereafter AGS), Simancas, Spain; Fernández to unknown, 9 May 1793, Aranjuez, exp. 1, leg. 6851, SGU; Sherry Johnson, *Social Transformation*, 79.
11. Antonio Fernández to unknown, 9 May 1793, Aranjuez, exp. 1, leg. 6851, SGU; Fernández, request for employment as lieutenant colonel of dragoons in Spanish army, 5 July 1793, exp. 1, leg. 6851, SGU.
12. [Marqués de la Torre], "Instruccion que ha de observar Don Antonio Fernandez. . . ." 16 January 1775, Havana, leg. 1220, PC, typescript copy in DSC; de la Torre to Julián de Arriaga, 22 January 1775, Havana, leg. 1220, PC, typescript copy in DSC.
13. [Marqués de la Torre], "Instruccion."
14. Joseph Alena, "Reglamento de Dragones de America. . . ." 22 July 1764, Havana, photocopy in *Archivo General de Indias, AGI, Documentos sobre las milicias de Cuba (1764–1785), Audiencia de Santo Domingo Leg.* 2078, vol. 106, LMC; Antonio Fernández to unknown, 9 May 1793, Aranjuez, exp. 1, leg. 6851, SGU; Fernández, request for employment; Victorina Guillén, widow's pension, 1800, exp. 105, leg. 7232, SGU; unknown to Gobernador de Santo Domingo, 3 December 1794, San Lorenzo, exp. 56, leg. 7151, SGU. Fernández's devotion to his post additionally was evident in his failure to request permission to return to Spain. By 1791, it had been thirty-one years since he had last visited the peninsula. See "Informe," 23 December 1791, Havana, exp. 56, leg. 7151, SGU.
15. [Marqués de la Torre], "Instruccion que ha de observar Don Antonio Fernández. . . ." 16 January 1775, Havana, leg. 1220, PC, typescript copy in DSC; Gelabert, "Relación del reconocimiento. . . ."
16. H. E. Friedländer, *Historia económica de Cuba* (Havana: Jesus Montero, 1944), 14:91; "Repartimiento de los empleados en la Factoría de Tabacos de la Habana. . . ." 30

August 1796, Havana, leg. 2001, SD; Marrero, *Cuba*, 9:216; Nuñez and Gonzalez, "Plan demostrativo."

17. "Repartimiento de los empleados"; "Instrucciones para el gobierno del ramo de tabacos. . . ." 1796, Madrid, leg. 2002, SD. "Most benign . . ." is from "Instrucciones."
18. Joaquín de Sola to unknown, 28 June 1768, Puerto Príncipe, doc. 30, leg. 1085, PC.
19. Fernando Muñoz to Antonio María Bucarely, 25 July 1768, Trinidad, leg. 1088, PC.
20. Luis de Echavarría to Antonio Rubio, 9 July 1765, Santiago de Cuba, no. 10, leg. 26, CG; Ramón Vuelta Flores to Antonio María de Bucarely, 23 July 1769, Puerto Príncipe, leg. 1074, PC; "Todos los Tavacos. . . ." 13 April 1774, Trinidad, leg. 1186, PC; Matias José Duran del Castillo et al. to Intendente, 27 November 1799, Mayarí, exp. 26, leg. 1114, IGH.
21. "Instrucciones para el gobierno del ramo de tabacos en la isla de Cuba," Madrid, 1796, legajo 2002, SD.
22. Ibid; Julián de Arriaga to Antonio María de Bucarely, 24 May 1771, Aranjuez, leg. 1140, PC.
23. "Repartimiento de los empleados"; Marrero, *Cuba*, 11:33; Rivero Muñiz, *Tabaco*, vols. 1–2.
24. Catherine Lugar, "The Portuguese Tobacco Trade and Tobacco Growers of Bahia in the Late Colonial Period," in *Essays Concerning the Socioeconomic History of Brazil and Portuguese India*, edited by Dauril Alden and Warren Dean, 26–70 (Gainesville: University Presses of Florida, 1977), 43–44.
25. "Instrucciones para el gobierno."
26. Duvon Clough Corbitt, "The Colonial Government of Cuba" (PhD diss., University of North Carolina, 1938), 29–30.
27. Kuethe, *Cuba, 1753–1815*, 26, 70–71; Leví Marrero, *Azúcar, ilustración y concencia (1763–1868) (IV)*, vol. 12 of *Cuba: Economía y sociedad* (Madrid: Editorial Playor, 1985), 314–16; Miguel de Altarriba, 22 January 1771, no. 1, leg. 120, Ultramar, AGI; Marqués de la Torre to Miguel de Múzquiz, 7 March 1776, Havana, leg. 1227, PC. The Marqués de Esquilache, Leopoldo di Grigorio, founded the institution of the intendancy, which was designed to improve the financial management of Havana's royal treasury and the supervision of military matters. Kuethe notes that although Arriaga signed the document finalizing its creation, it was Esquilache who was responsible for its conception. Marrero inaccurately states the period of Rapún's service as 1773–1778.
28. Marqués de la Torre to Miguel de Múzquiz, 7 March 1776, Havana, leg. 1227, PC, AGI.
29. Leví Marrero, *Azucar, ilustración y conciencia (1763–1868) (II)*, vol. 10 of *Cuba: Economía y sociedad* (Madrid: Editorial Playor, 1984), 114; Kuethe, *Cuba*, 52; Sherry Johnson, *Social Transformation*, 155; Marqués de la Torre to Miguel de Múzquiz, 7 March 1776, Havana, leg. 1227, PC; Juan Ignacio de Urriza to de la Torre, 17 March 1776, Havana, leg. 1153, PC.
30. "The issue to destroy . . ." is from James A. Lewis, "Anglo-American Entrepreneurs in Havana: The Background and Significance of the Expulsion of 1784–1785," in *The North American Role in the Spanish Imperial Economy, 1760–1819*, edited by Jacques A. Barbier and Allan J. Kuethe (Manchester, UK: Manchester University

Press, 1984), 120. For the intendant's correspondence to the captain general, see Juan Ignacio de Urriza to Diego José Navarro, 21 August 1779, Havana, leg. 1238, PC; Urriza to Navarro, 23 August 1779, Havana, leg. 1238, PC; Urriza to Navarro, 23 August 1779, Havana, leg. 1238, PC; Urriza to Navarro, 24 August 1779, Havana, leg. 1238, PC; Urriza to Navarro, 18 September 1779, Havana, leg. 1238, PC; Urriza to Navarro, 20 September 1779, Havana, leg. 1238, PC; Urriza to Navarro, 21 September 1779, Havana, leg. 1238, PC; Urriza to Navarro, 22 September 1779, Havana, leg. 1238, PC; Urriza to Navarro, 30 September 1779, Havana, leg. 1238, PC; Urriza to Navarro, 2 October 1779, Havana, leg. 1238, PC; Urriza to Navarro, 4 October 1779, Havana, leg. 1238, PC; Urriza to Navarro, 6 October 1779, Havana, leg. 1238, PC; Urriza to Navarro, 26 October 1779, Havana, leg. 1238, PC; Urriza to Navarro, 29 October 1779, Havana, leg. 1238, PC; Urriza to Navarro, 8 November 1779, Havana, leg. 1238, PC; Urriza to Navarro, 9 November 1779, Havana, leg. 1238, PC; and Urriza to Navarro, 13 November 1779, Havana, leg. 1238, PC.

31. Kuethe, *Cuba*, 158. Gómez Robaud's connection with Beltrán de Santa Cruz y Cárdenas was extremely valuable, since the count was extremely well-placed in Cuban society and had particular access to the Spanish Crown. Married to María Teresa Montalvo y O'Farill, daughter of one of the island's richest men, the first Conde de Casa Montalvo, Beltrán de Santa Cruz y Cárdenas was related (by blood and by her grandfather's second marriage) to at least four other members of Cuba's titled families: the Marqués de Real Socorro, the Conde de Casa Bayona, the Conde de Buena Vista, and the Marqués de Casa Calvo. He lived for a while in Spain to further his personal ambitions, and became a "*gentilhombre de cámara* (gentleman of the Royal Bedchamber) with entrance in 1792." Allan J. Kuethe, "*Los Llorones Cubanos*: The Socio-military Basis of Commercial Privilege in the American Trade," in *The North American Role in the Spanish Imperial Economy, 1760–1819*, edited by Jacques Barbier and Allan J. Kuethe (Manchester: Manchester University Press, 1984), 152. By the mid-1790s, Beltrán de Santa Cruz y Cárdenas had befriended Manuel Godoy, a highly decorated confidante of Carlos IV and Queen María Luisa's lover. Beltrán de Santa Cruz y Cárdenas was granted the title of Conde de Santa Cruz y Mopox while in Spain, and inherited the title of Condado de San Juan y Jaruco in 1804. Kuethe, *Cuba*, 120, 152–54, 158; John Lynch, *Bourbon Spain, 1700–1808* (Oxford: Blackwell, 1989), 385–87. On the count's involvement with the tobacco industry, see Chapter 3.

32. Rafael Gómez Robaud, "Ynforme. . . .," 25 January 1819, Valencia, typescript in *Archivo General de Indias, Cuba, 1720–1819*, 25:398–485, LMC. Gómez Robaud's support of the island's tobacco farmers placed him in opposition to monopoly advisor Francisco Arango y Parreño. This feud, which grew to include former captain general Luis de las Casas, is detailed in Sherry Johnson, *Social Transformation*, 169–73, and Sherry Johnson, "From Authority to Impotence: Arango's Adversaries and Their Fall from Power during the Constitutional Period (1808–1823)," in *Francisco Arango y la invención de la Cuba azucarera*, edited by María Dolores González-Ripoll and Izaskun Álvarez Cuartero (Salamanca, Spain: Ediciones Universidad Salamanca, 2009), 193–211, Arango's well-placed friends are noted in José A. Piqueras, "Los amigos de Arango en la Corte de Carlos IV," in González-Ripoll and Álvarez Cuartero, *Francisco Arango*, 151–66.

33. Sherry Johnson, *Social Transformation*, 172; "Instrucciones para el gobierno."
34. Employment record, Braulio de Vivanco, circa 31 August 1817, Real Factoría de Tabacos, Havana, leg. 2019, SD; "Repartimiento de los empleados"; employment record, Braulio de Vivanco, end of October 1821, Havana, leg. 252, Ultramar, AGI; Felipe de la Puente to unknown, 19 February 1824, Madrid, leg. 252, Ultramar, AGI.
35. Employment record, Braulio de Vivanco, circa 31 August 1817; José de Gálvez, license for Antonio Aguilar de Cela.
36. Sherry Johnson, *Social Transformation*, 58. See Sherry Johnson, *Social Transformation*, 39–120, for the impact and role of Spanish soldiers and retired military in colonial Cuba.
37. Among the histories of this period are Louis A. Pérez Jr., *Cuba between Empires, 1878–1902* (Pittsburgh: University of Pittsburgh Press, 1998); Pérez, *Cuba: Between Reform and Revolution*; Ada Ferrer, *Insurgent Cuba: Race, Nation, and Revolution, 1868–1898* (Chapel Hill: University of North Carolina Press, 1999); Aline Helg, *Our Rightful Share: The Afro-Cuban Struggle for Equality, 1886–1912* (Chapel Hill: University of North Carolina Press, 1995); Rosalie Schwartz, *Lawless Liberators: Political Banditry and Cuban Independence* (Durham, NC: Duke University Press, 1989); Philip Sheldon Foner, *Antonio Maceo: The "Bronze Titan" of Cuba's Struggle for Independence* (New York: Monthly Review Press, 1977).
38. Gabriel Santos, "Lista que comprehende los negros esclavos de Rl Compania de esta Ciudad e Ysla, con destino en el Ramo de Tavacos, y distinction de sus nombres, naziones, Fihaziones, tachas y Edades, los mismos que constant del Ymbentario gral de Existencias del dia treinta y uno de octubre proximo pasado, formando con esta fha. a saver," 17 February 1753, Real Factoría de Tabacos, Havana, leg. 1002, Ultramar, AGI. For more on slaves in Cuba's tobacco industry, see Vincent Sanz Rozalén, "Los negros del Rey: Tabaco y esclavitud en Cuba a comienzos del siglo XIX," in *Trabajo libre y coactivo en sociedades de plantación*, edited by José A. Piqueras (Madrid: Siglo XXI Editores, 2009), 151–76.
39. Humboldt, *Island of Cuba*, 287, cited in Herbert Klein, *Slavery in the Americas*, 159; Capitán del Partido de Jesús, María, y José, "Padron general," 30 June 1791, Jesús, María, y José, leg. 1471, PC; "Noticia del estado de posesiones y enceres de la Factoría de Tabacos de la Habana en fin de deciembre de 1812 del numo 40 de las cuentas del propio año, con expresion de los valores," December 1812, leg. 2001, AGI.
40. Robert J. Shafer, "Ideas and Work of the Colonial Economic Societies, 1781–1820," *Revista de Historia de América* 44 (December 1957): 348; "Instrucciones para el Gobierno"; Gómez Roubaud, "Ynforme"; Ignacio González to Superintendente Director General Into., 11 February 1811, Havana, exp. 12, leg. 973, IGH.
41. Junta, 15 July 1802, Havana, leg. 235, Ultramar, AGI; Antonio del Valle Hernández to Francisco Arango y Parreno, 14 April 1810, Havana, leg. 245, Ultramar, AGI.
42. Astigarrota, "Informe sobre resguardo de a cavallo"; Urriza, "Nomina."
43. The areas in which tobacco could be planted were strictly controlled in Mexico and Colombia, for example. Although "the King indicated that he intended to abolish all planting in the colony . . . his 'royal clemency' inclined him to permit growing in the jurisdictions of Cordoba, Orizaba, and Teusitlan" in the modern-day state of Veracruz (McWaters, "Royal Tobacco Monopoly," 32). The first two

locations were chosen in part because the mountainous geography of the area allowed for easy guarding of the crop, and therefore less possibility of contraband. Despite the order limiting tobacco growing to only three areas, Jalapa was added by the *junta* on 4 September 1765, in part because of pleas from the *alcalde mayor* (district royal magistrate) and local merchants who argued that tobacco had been planted there (in their words) "from time immemorial" and because it was the area's only significant agricultural crop. By 1769, however, reports of contraband outside Cordoba convinced the director of Mexico's tobacco monopoly of the need to limit the growing area (McWaters, "Royal Tobacco Monopoly," 61–62, 71–73). After Colombia's tobacco monopoly was reorganized in 1778, tobacco growing was restricted to "four small and relatively compact areas that were easy to police" (Harrison, "Evolution," 165).

44. Ramires de Arellano, "Estado que manifiesta los Tabacos"; Marqués de la Torre to Martín de Asanza, 23 September 1774, Havana, leg. 1179, PC; Sherry Johnson, *Climate and Catastrophe*. A sampling of prohibitions against increasing the number of *vegas* includes Antonio Fernández to de la Torre, 21 April 1775, Filipinas, leg. 1180, PC; Joaquín de Sola to de la Torre, 2 August 1775, Puerto Príncipe, doc. 83, leg. 1173, PC; de la Torre to don Juan Mancebo y Castañeda, 23 July 1776, Havana, leg. 1179, PC. Farmers in one area were told that they could keep the tobacco plants already in the ground. See de la Torre to Ambrosio Pérez Baes, 12 May 1775, Havana, leg. 1193, AGI.

45. José Domingo Morejon to Luis de Unzaga, 26 June 1784, Batabanó, doc. 97, leg. 1351, PC; Marrero, *Cuba*, 9:217; Sherry Johnson, *Social Transformation*, 54. In 1817, tobacco was forever linked to Güines when a royal decree approved the town's official coat of arms, which depicted a tobacco plant, not sugar, on either side and below the *villa*'s patron saint, Saint Julián. Moreno Fraginals, Hugh Thomas, and more recently Alfonso W. Quiroz emphasize the role that sugar played in Güines by the early 1800s. See Moreno Fraginals, *Sugarmill*, 19–20, 22–23; Hugh Thomas, *Cuba, or the Pursuit of Freedom*, updated ed. (New York: Da Capo Press, 1998), 77; and Alfonso W. Quiroz, "Cambio en Güines en la era de Arango y Humboldt," in González-Ripoll and Álvarez Cuartero, *Francisco Arango*, 229–43.

46. Francisco Cagigal de la Vega to Alonso Arcos y Moreno, 21 August 1750, Havana, order 169, leg. 5, CG; Cagigal de la Vega to Arcos y Moreno, 4 September 1750, Havana, order 167, leg. 5, CG; Julián de Arriaga to Antonio María de Bucarely, 24 January 1771, El Pardo, leg. 1156, PC. The August 1750 letter makes reference to similar requests made in previous correspondence dated 20 May, 26 July, and 2 and 27 September 1749. The French also were likewise concerned with protecting their monopoly's sole rights to their national tobacco industry. Anyone selling tobacco outside of the official French monopoly was subject to penalties, which by July 1681 included three years' banishment and a fine for a second offense, while a third offense sent the guilty party to the pillory, followed by permanent banishment (Price, *France and the Chesapeake*, 11–12, 17–19, 124–26).

47. Marqués de Casa Cagigal to Antonio María de Bucarely, 15 May 1766, Santiago de Cuba, doc. 12, leg. 1051, PC; Andres José Carillo to Bucarely, 27 July 1768, San José de Remedios, leg. 1088, PC; Julián de Arriaga to Bucarely, 22 September 1770, San Ildefonso, folio 121, leg. 1134, PC; Arriaga to Bucarely, 24 January 1771, El Pardo, leg. 1156, PC; Marqués de la Torre to Alcaldes Ordinarios de la Ciudad

de San Carlos de Matanzas, 5 July 1772, Havana, doc. 6, leg. 1187, PC; Antonio de la Torre to Marqués de la Torre, 24 September 1774, San Juan de Remedios, doc. 68, leg. 1185, PC; Manuel Coimbra to Luis de Unzaga, 8 August 1783, Havana, doc. 1, leg. 1351, PC; Alejandro Ramirez to Secretario de Estado y del Despachado de Hacienda de España, 10 October 1817, Havana, leg. 2019, SD; Bucarely to Sr. Bentura Doral, 31 March 1766, Puerto Príncipe, leg. 1077, PC; Bucarely to unknown, 17 May 1771, Havana, doc. 42, leg. 1077, PC; notice, 10 November 1782, Filipinas, leg. 1317, PC.

48. Ramirez de Arellano, "Estado que manifiesta los distintos precios." The seven classes of tobacco were *largo*, *corto*, *basura limpia*, *desecho limpio*, *desechito*, *libras*, and *infuriado*.

49. The Spanish Crown in 1802 earned 643,032 *pesos* on tobacco purchased from the island's *vegueros* and resold through the monopoly. Manuel Ramírez de Arellano, "Demostración del number de Arrobas de Tabaco vendidos. . . ." 4 August 1804, Real Factoría de Tabacos, Havana, table 12, signature 9454, leg. 209, RC; Ramirez de Arellano, "Estado que manifiesta los distintos precios"; "Copia del Estado remitido por el Director General de la Factoría con carta de 14 de July de 814. . . ." [after 1813], leg. 2001, SD.

50. For a sample of correspondence on this issue, see Antonio de la Torre to Marqués de la Torre, 15 January 1776, San Juan de los Remedios, leg. 1185, PC; Vicente Manuel de Cespedes to Diego José Navarro, 4 May 1781, Santiago de Cuba, leg. 1231, PC; and "Informe sobre las disposiciones tomadas con los tabacos," 29 April 1815, leg. 2001, SD. The monopoly's chronic shortage of funds also affected its transactions with growers in other parts of the Spanish empire: in November 1769, Cuba's governor warned the viceroy of New Spain that there were no funds to pay New Spain's *vegueros* for the previous harvest. Money to pay the farmers did not arrive until August 1770. See Laura Náter, "El negocio colonial de tabaco en el Imperio Español," unpublished manuscript, 2001, 33.

51. Manuel del Portillo to Marqués de la Torre, 13 June 1775, Sancti Spíritus, doc. 79, leg. 1185, PC; Pedro Suarez de Urbina to Intendente del Ramo de Tabaco, 18 October 1813, Santiago de Cuba, leg. 2001, SD; José María de la Torre to Miguel Cayetano Soler, 5 February 1805, Havana, leg. 2018, SD; Rafael Gómez Robaud to Francisco de Saavedra, 9 January 1810, Havana, leg. 2001, SD. De la Torre's quote reads, "Con esperanzas de mejor trato y consideraciones en los succesvio . . ."

52. Marqués de la Torre to Mateo de Echavarría y Elguerra, 11 August 1775, Havana, leg. 1156, PC; de la Torre to Echavarría, 28 October 1775, Havana, leg. 1156, PC; de la Torre to Echavarría, 5 January 1776, Havana, leg. 1156, PC.

53. Corbitt, "Colonial Government of Cuba," 65–66; Governor of Havana to the Marqués de Sonora, José de Gálvez, 17 November 1786, *carta reservada* 12, signature 15, leg. 1034, IGH; José de Gálvez to José Pablo Valiente, 7 March 1787, El Pardo, no. 15, leg. 1034, IGH.

54. Royal order, 7 November 1803, San Lorenzo, exp. 12, leg. 96, IGH.

CHAPTER 6

1. Lance Grahn, *The Political Economy of Smuggling: Regional Informal Economies in Early Bourbon New Granada*, Dellplain Latin American Studies, no. 35, edited by David R. Robinson (Boulder, CO: Westview Press, 1997), 3, 5–6. The connections that drew

disparate social and economic classes together in Cuba through the production of tobacco were discussed in detail in Chapter 3.

2. A tiny sampling of works that have addressed contraband includes Vera Lee Brown, "The South Sea Company and Contraband Trade," *American Historical Review* 31, no. 4 (July 1926): 662–78; and Robert C. Nash, "The English and Scottish Tobacco Trade in the 17th and 18th Centuries: Legal and Illegal Trade," *Economic History Review* 35, no. 3 (1982): 354–72.
3. Braulio de Vivanco to Señores Directores Generales de Rentas, 15 April 1818, Havana, leg. 2019, SD.
4. Governor Juan Francisco Güemes de Horcasitas to the king, 10 December 1741, Havana, leg. 386, SD, cited in Marrero, *Cuba*, 7:118; Violet Barbour, "Privateers and Pirates of the West Indies," *American Historical Review* 16, no. 3 (April 1911), 531; Kenneth R. Andrews, *The Spanish Caribbean: Trade and Plunder, 1530–1630* (New Haven, CT: Yale University Press, 1978), 167; McNeill, *Atlantic Empires*, 198; Klooster, "Illicit Riches," 72.
5. Kenneth R. Andrews, *Spanish Caribbean*, 156; Kenneth R. Andrews, *Elizabethan Privateering: English Privateering during the Spanish War, 1585–1603* (Cambridge: Cambridge University Press, 1964), 191; Jeremy Cohen, "Smuggling in Eighteenth-Century Venezuela: The Role of the *Práctico*," paper presented at the meeting of the Latin American Studies Association (Dallas, March 2003), 3.
6. Klooster, "Illicit Riches," 72; Louis A. Pérez Jr., *Cuba: Between Reform and Revolution*, 4th ed., 40.
7. Unknown to Francisco Cagigal de la Vega, 11 September 1747, order 5, leg. 5, CG; Francisco Ródriguez y Olivera to Lorenzo de Madariaga, 1 September 1757, Bayamo, order 47, leg. 7, CG. Additional examples of the presence of foreign ships during this period intent on contraband are documented in Cagigal to unknown, 16 March 1756, Santiago de Cuba, order 170, leg. 7, CG; José Antonio de Quiroga to Madariaga, 13 December 1757, Juraguacito, order 17, leg. 7, CG; and McNeill, *Atlantic Empires*, 184. McNeill notes, however, that the death penalty was rarely enforced. Hubert H. S. Aimes further explains that the contraband trade "centered around Cuban tobacco" and that the "active campaign against the smugglers . . . led to the war of 1740 between Spain and Great Britain" (*Slavery in Cuba*, 22).
8. Quoted in Frances Armytage, *The Free Port System in the British West Indies: A Study in Commercial Policy, 1766–1822* (London: Longmans, Green, 1953), 20–21.
9. Quoted in Celestino Andrés Araúz Monfante, "La acción ilegal de los holandeses en el Caribe y su impacto en las Antillas y Puerto Rico durante la primera mitad del siglo XVIII," *Revista/Review Interamericana* 14, nos. 1–4 (Spring/Winter 1984): 73–74.
10. See, for example, legs. 1051, 1078–79, and 1463A–B, PC; and legs. 1131, 1134–36, and 1138–39, SD. Louis A. Pérez Jr. notes that during the ten-month period of the British occupation, "an estimated 10,000 slaves were introduced to Havana—as many slaves as would have normally entered in ten years" (*Cuba: Between Reform and Revolution*, 58).
11. For a small sampling of works that demonstrate the depth and breadth of the contraband tobacco trade in the Atlantic world, see Coutts, "Boom and Bust"; Joyce Lorimer, "The English Contraband Tobacco Trade in Trinidad and Guiana, 1590–1617," in *The Westward Enterprise: English Activities in Ireland, the Atlantic, and*

America, 1480–1650, edited by K. R. Andrews, N. P. Canny, and P. E. H. Hair (Liverpool: Liverpool University Press, 1978), 124–50; Arthur Pierce Middleton, "The Chesapeake Convoy System, 1662–1763," *William and Mary Quarterly*, 3rd series, 3, no. 2 (April 1946): 182–207; and Thomas L. Whigham, "Agriculture and the Upper Plata: The Tobacco Trade, 1780–1865," *Business History Review* 59, no. 4 (Winter 1985): 573–75.

12. Quoted in Kenneth R. Andrews, *Spanish Caribbean*, 227.
13. Neville Williams, *Contraband Cargoes: Seven Centuries of Smuggling* (Camden, CN: Shoe String Press, 1961), 9, 124; B. W. E. Alford, *W. D. and H. O. Wills and the Development of the U.K. Tobacco Industry, 1786–1965* (London: Methuen, 1973), 5; C. M. MacInnes, *Early English Tobacco Trade* (London: Kegan Paul, Trench, Trubner, and Company, 1926), 34, 53, 150; Carl A. Hanson, "Monopoly and Contraband in the Portuguese Tobacco Trade, 1624–1702," *Luso-Brazilian Review* 19, no. 2 (Winter 1982): 154.
14. Rivero Muñiz, *Tabaco*, 1:38; Pérez, *Cuba: Between Reform and Revolution*, 55; McNeill, *Atlantic Empires*, 155; Aimes, *Slavery in Cuba*, 22. For *bandos*, see, for example, Andres José Carrillo to Antonio María de Bucarely, 27 July 1768, San Juan de Remedios, leg. 1088, PC; Julián de Arriaga to Bucarely, 24 January 1771, El Pardo, leg. 1156, PC; Joaquín de Sola to Marqués de la Torre, 9 July 1775, Puerto Príncipe, doc. 87, leg. 1173, PC; and Antonio Daza Maldonado to Superintendente Director General, 25 November 1809, Havana, no. 59, leg. 1060, IGH. These types of proclamations date back to at least the 1750s. See Francisco Cagigal de la Vega to Alonso Arcos y Moreno, 21 August 1750, Havana, doc. 169, leg. 5, CG; and Cagigal de la Vega to Arcos y Moreno, 2 June 1754, Havana, doc. 49, leg. 6, CG. For tobacco from western Cuba, see Rafael Gómez Robaud to José de la Torre, 24 October 1794, Havana, leg. 2002, SD.
15. Sherry Johnson, *Social Transformation*, 77–78; Juan Gemir y Lleonart to Antonio María de Bucarely, 26 July 1768, Bayamo, leg. 1080, PC; Joaquín de Sola to Bucarely, 28 June 1768, Puerto Príncipe, leg. 1085, PC; Fernando Muñoz to Bucarely, 25 July 1768, Trinidad, leg. 1088, PC.
16. Consejo de Indias, 17 February 1753, Madrid, leg. 1131, SD; Consejo de Indias, 13 March 1753, Madrid, leg. 1131, SD; Consejo de Indias, 24 March 1753, Madrid, leg. 1131, SD; Consejo de Indias, 22 October 1767, Madrid, leg. 1135, SD; Consejo de Indias, 13 November 1767, Madrid, leg. 1135, SD; Consejo de Indias, 19 July 1769, leg. 1136, SD.
17. Julián de Arriaga to Antonio María de Bucarely, 24 January 1771, El Pardo, leg. 1134, PC. In the case of Antonio Alfonso, it was suggested that the fine be raised to one hundred ducats. See Marqués de la Torre to Nicolás José Rapún, 16 February 1774, Havana, folio 584, doc. 530, leg. 1154, PC.
18. Marqués de la Torre to Miguel de Altarriva, 26 November 1772, Havana, folio 223, doc. 266, leg. 1154, PC; de la Torre to Mateo de Echavarría y Elguerra, 16 February 1774, Havana, leg. 1156, PC; de la Torre to Nicolás José Rapún, 16 February 1774, Havana, folio 584, doc. 530, leg. 1154, PC; de la Torre to Rapún, 6 April 1774, Havana, folio 613, doc. 553, leg. 1154, PC. By April, it had been determined that the tobacco seized from Alfonso's *estancia* belonged to Miguel Hernández.
19. Marqués de la Torre to Martín Xavier de Echavarría, 7 February 1775, Havana, leg. 1156, PC. For other seizures during the early 1770s, see Adrian José de Armas to de la Torre, 5 October 1774, Horcón, folio 23, leg. 1194, PC; Pablo Gutierrez

Muxica to de la Torre, 15 October 1774, San Juan de Remedios, doc. 50, leg. 1185, PC; and de la Torre to Echavarría, 22 February 1776, leg. 1156, PC. For tobacco production statistics, see Ramirez de Arellano, "Estado que manifiesta los Tabacos." Tobacco production for the Guane district was admittedly low this particular year, at only 442 *arrobas* 4 *libras* (or about 11,054 pounds), but it was followed by a ten-year (1765–1775) high of 4,480 *arrobas* 19 *libras* (or approximately 112,019 pounds) in 1773.

20. Consejo de las Indias, 19 July 1769, leg. 1136, SD; Alejandro Mora, *procesos*, doc. 68, leg. 7135, Guerra Moderna (hereafter GM), AGS; Lorenzo Cendra, *procesos*, destino al Regimiento Fijo del reo Lorenzo Cendra, 1787–1789, doc. 3, leg. 7149, GM; Juan Gutierrez to Antonio María de Bucarely, 31 July 1768, doc. 28, leg. 1088, PC. Beginning on 22 May 1783, sentences in the Philippines, where illegal trade in tobacco was particularly acute, were stiffened; the new penalties included subjecting contrabanders to eight years of hard labor in the royal arsenal (de Jesus, *Tobacco Monopoly*, 48).
21. Marqués de la Torre to Martín Xavier de Echavarría, 22 February 1776, Havana, leg. 1156, PC; Diego José Navarro to Juan Ignacio de Urriza, 6 August 1779, Havana, leg. 1238, PC; Astigarrota, "Informe sobre resguardo de a cavallo"; Garcia Alayeto, "Instrucciones para el resguardo montado." There is at least one instance where the informant was rewarded with half of the tobacco's value. See de la Torre to Echavarría, 17 September 1776, Havana, leg. 1156, PC. Informants were also rewarded in Mexico. See McWaters, "Royal Tobacco Monopoly," 37.
22. For tobacco seized, see Juan Ignacio de Urriza to Diego José Navarro, 18 September 1779, Havana, leg. 1238, PC; Urriza to Navarro, 22 September 1779, Havana, leg. 1238, PC; Urriza to Navarro, 2 October 1779, Havana, leg. 1238, PC; Urriza to Navarro, 26 October 1779, Havana, leg. 1238, PC; Urriza to Navarro, 8 November 1779, Havana, leg. 1238, PC; Urriza to Navarro, 9 November 1779, Havana, leg. 1238, PC; and Urriza to Navarro, 13 November 1779, Havana, leg. 1238, PC. For a list of the original *resguardo montado* members, see Urriza, "Nomina."
23. Astigarrota, "Informe sobre resguardo de a cavallo."
24. José de la Torre to Luis de Unzaga, 1 August 1784, Pinar del Río, leg. 1350, PC; Bartolomé Castellanos to Administrador de la Real Factoría de Tavacos, 12 April 1800, Havana, doc. 1, leg. 898, IGH; Manuel Trensa to Superintendente Director General, 26 September 1816, Havana, doc. 8, leg. 1043, IGH; José González to Intendente del Exercito, 28 September 1816, Havana, doc. 8, leg. 1043, IGH; Agustín Alvarez to Rafael Gómez Roubaud, 30 September 1816, Havana, doc. 8, leg. 1043, IGH.
25. Bartolomé Castellanos to Administración de la Real Factoría, 12 April 1800, Havana, doc. 1, leg. 898, IGH; Licenciado Juan de Arechaga, testimony, 23 April 1800, doc. 1, leg. 898, IGH; "Delito oficial contra Juan de Acosta, veguero del partido de San Juan y Martínez sobre averiguar el destino que oio a su cosecha de tabaco el año anterior," 1804, doc. 26, leg. 1003, IGH.
26. Consejo de Indias, 22 October 1767, Madrid, leg. 1135, SD; Mateo de Echavarría y Elguerra to Diego José Navarro, 27 January 1778, Santiago de Cuba, leg. 1231, PC.
27. Antonio María de Bucarely to unknown, 20 October 1768, Havana, doc. 27, leg. 1077, PC; Dick Cluster and Rafael Hernández, *The History of Havana* (New York: Palgrave Macmillan, 2006), xv; Le Riverend Brusone, *La Habana, espacio y vida*, 138;

Bucarely to unknown, 20 October 1768, Havana, doc. 27, leg. 1077, PC; Manuel García to Marqués de la Torre, 8 July 1776, Havana, leg. 1156, PC; Martín Xavier de Echavarría to de la Torre, 25 August 1776, leg. 1156, PC; Echavarría to de la Torre, 31 August 1776, Havana, leg. 1156, PC.

28. Clara Elena Suárez Argüello, *Camino real y carrera larga: La arriería en la Nueva España durante el siglo XVIII* (Mexico City: Centro de Investagiciones y Estudios Superiores en Antropología Social, Ediciones de la Casa Chata, 1997), 71–72, 112; Esteban Rodríguez del Pino to Marqués de la Torre, 19 January 1772, San Juan y Santiago, leg. 1167, PC; Juan Ignacio de Urriza to Diego José Navarro, 4 October 1779, Havana, leg. 1238, PC. *Arrieros* elsewhere engaged in illegal activities and took advantage of their positions to attempt to steal tobacco. In Mexico, Indian tributary and mule pack driver José Esteban was sentenced in 1798 to one and a half years in prison for stealing tobacco while transporting it from Córdoba to Mexico City. See Deans-Smith, *Bureaucrats, Planters, and Workers*, 32. In Costa Rica, corn was carried on the top of the load with tobacco concealed underneath. See Fallas, *La factoría*, 174.

29. Francisco Cagigal de la Vega to Alonso Arcos y Moreno, 30 April 1751, Havana, order 239, leg. 7, CG; Cagigal de la Vega to Arcos y Moreno, 30 April 1754, Havana, order 207, leg. 7, CG; Cagigal de la Vega to Arcos y Moreno, 30 April 1751, Havana, order 141, leg. 6, CG; Cagigal de la Vega to Lorenzo de Madariaga, 20 July 1755, Havana, order 89, leg. 7, CG; Madariaga to Cagigal de la Vega, 10 September 1755, Santiago de Cuba, order 255, leg. 7, CG; Arcos y Moreno to Cagigal de la Vega, 7 June 1752, Santiago de Cuba, order 234, leg. 6, CG; Mariano Lacueba, *relación*, 5 May 1777, Havana, folio 1260, leg. 1153, PC.

30. Urriza, "Nomina"; Astigarrota, "Informe sobre resguardo de a cavallo"; Juan Ignacio de Urriza to Diego José Navarro, 22 September 1779, Havana, leg. 1238, PC; Urriza to Navarro, 2 October 1779, leg. 1238, PC. To put the salary in context, the amount earned by Sorrivas was significantly less than the 2,400 *pesos* a colonel in Havana's regiment received in 1759, but more than that of a soldier in Havana's garrison. See Marrero, *Cuba*, 7:92.

31. Juan Ignacio de Urriza to Diego José Navarro, 26 October 1779, Havana, leg. 1238, PC; Urriza to Navarro, 29 October 1779, Havana, leg. 1238, PC; Urriza to Navarro, 8 November 1779, Havana, leg. 1238, PC.

32. Francisco Antonio de Astigarrota to José Pablo Valiente, 29 October 1787, Havana, 29 October 1787, no. 15, leg. 1034, IGH; Astigarrota, "Informe sobre resguardo de a cavallo"; José Francisco Grillo et al., testimony, 25 October 1779, Govea, leg. 1267, PC. While Serrano's immediate fate is unclear, by 1787 he was no longer a member of the *resguardo*.

33. Alfonso W. Quiroz, "Implicit Costs of Empire: Bureaucratic Corruption in Nineteenth-Century Cuba," *Journal of Latin American Studies* 35, no. 3 (August 2003): 473–511. Cuba's *resguardo* was not the only colonial Spanish entity suspected of collusion. Susan Deans-Smith notes that Mexican *resguardo* members were believed to be suspect, "although no documented cases came to light" (*Bureaucrats, Planters, and Workers*, 91).

34. Marqués de la Torre to Señora doña Justina Isabel Laso, 15 May 1773, Havana, leg. 1179, PC.

35. McNeill, *Atlantic Empires*, 184; Quiroz, "Implicit Costs," 510; Fallas, *La factoría*, 65; Deans-Smith, *Bureaucrats, Planters, and Workers*, 47–49. G. Earl Sanders argues for

a more moderate view of these bureaucrats; see "Counter-Contraband in Spanish America: Handicaps of the Governors in the Indies," *Americas* 34, no. 1 (July 1977): 59. Portuguese monks and nuns in the 1670s not only hid and sold tobacco within monastery and convent walls, but also processed it into manufactured products (Hanson, "Monopoly and Contraband," 154–55), while in New Spain, the Bishop of Oaxaca "sold contraband tobacco under the protection of ecclesiastical immunity" (Deans-Smith, *Bureaucrats, Planters, and Workers*, 22).

36. Butler, "Power in Smoke," 114–16.
37. José de Gálvez to José Pablo Valiente, 7 March 1787, El Pardo, no. 15, leg. 1034, IGH. See Kuethe, *Cuba, 1753–1815*, 94–95, 106–12, for the Algiers and Pensacola expeditions.
38. Francisco Antonio de Astigarrota to José Pablo Valiente, 29 October 1787, Havana, exp. 15, leg. 1034, IGH; Fray Rogerio Navarro to Barbara Almirante, receipt, Convento de Nuestra Señora de la Merced, 15 February 1800, Havana, exp. 2, leg. 898, IGH; "Cuenta y relación firmada que yo Doña Barbara Almirante, como viuda y albacea terredora de bienes de los que quedaron por fallecimiento del Tte de Guarda Mayor Don Ramón Bretos he formado de los que han enterado en mi poder, y corrcerbo a disposición del Señor Intendente General de Exercito y Real Hacienda Juez de su testa y es como sigue," 4 August 1800, Havana, exp. 2, leg. 898, IGH; Licenciado Francisco Enziaro de Abreau, account record, Barbara Almirante, 3 February 1801, exp. 2, leg. 898, IGH. Bretos left a widow, Barbara Almirante, and five minor children, Domingo, Antonio, María Luisa, Francisca, and María de Jesús. An accounting of his assets included houses and a lot located in the Barrio Jesús María. Bretos was buried on 16 February in the habit of Nuestra Señora de la Merced, for which his wife paid sixteen pesos. The service was held in the parish church of Santo Ángel del Custodio, located in the city's northwestern corner, at a cost of 248 *pesos*. See "Memoria de entierro de Don Ramón Bretos hecho en la Iglesia Auxro de Santo Angel Custodio en 16 de February de 1800," exp. 2, leg. 898, IGH.
39. Ignacio Álvarez, employment record of Tomás Sánchez, 31 December 1808, Real Factoría de Tabacos, Havana, leg. 243, Ultramar, AGI; Álvarez, employment record of don Francisco Ximénez, 31 December 1808, Real Factoría de Tabacos, Havana, leg. 243, Ultramar, AGI; Álvarez, employment record of don Francisco Cabrera y Martínez, 31 December 1808, Real Factoría de Tabacos, Havana, leg. 243, Ultramar, AGI; Álvarez, employment record of don José María Pacot, 31 December 1808, Real Factoría de Tabacos, Havana, leg. 243, Ultramar, AGI; Álvarez, employment record of don Domingo Menendez San Pedro, 31 December 1808, Real Factoría de Tabacos, Havana, leg. 243, Ultramar, AGI.
40. James A. Lewis, "Anglo-American Entrepreneurs," 120. Lewis also notes that Urriza was responsible for pursuing the contraband case against Francisco de Miranda.
41. Sherry Johnson, *Social Transformation*, 81.
42. Francisco Cagigal de la Vega to Alonso Arcos y Moreno, 12 February 1754, Havana, leg. 6, CG; Cagigal de la Vega to Arcos y Moreno, 2 June 1754, Havana, doc. 49, leg. 6, CG; José de la Torre to Rafael Gómez Roubaud, 19 October 1804, Havana, leg. 2018, SD. Illegal tampering was common in the United Kingdom's tobacco industry during the 1830s, and although authorities were aware of this

practice, it was sometimes difficult to determine when a crime had taken place. This was particularly the case when more sophisticated methods were used, like those involving saccharin-based solutions, which were virtually impossible to detect without chemical analysis (Alford, *W. D. and H. O. Wills*, 76).

43. Antonio María de Bucarely to Julián de Arriaga, 9 October 1771, Havana, doc. 79, leg. 1140, PC; Miguel de Múzquiz to Marqués de la Torre, 24 June 1774, Aranjuez, leg. 1227, PC.

44. Miguel de Múzquiz to Marqués de la Torre, 1 September 1772, San Ildefonso, leg. 1227, PC; Múzquiz to de la Torre, 24 October 1772, San Lorenzo, folios 914–17, leg. 1227, PC; Múzquiz to de la Torre, 22 March 1773, folio 791, leg. 1227, PC; Múzquiz to de la Torre, 22 March 1773, folio 795, leg. 1227, PC; de la Torre to Juan José Claro, 11 June 1773, Havana, leg. 1181, PC; Vicente Carraso to de la Torre, 21 July 1773, Seville, folio 820, leg. 1227, PC; Múzquiz to de la Torre, 24 December 1773, Madrid, folios 839–40, leg. 1227, PC.

45. Marqués de la Torre to Miguel de Múzquiz, 8 October 1773, Havana, leg. 1227, PC; Martín Xavier de Echavarría to de la Torre, 8 October 1773, Havana, folios 1029–34, leg. 1227, PC; Múzquiz to de la Torre, 6 September 1774, San Ildefonso, doc. 43, folios 866–67, leg. 1227, PC; Múzquiz to de la Torre, 16 September 1774, San Ildefonso, doc. 45, folios 870–71, leg. 1227, PC; Múzquiz to de la Torre, 6 September 1774, San Ildefonso, doc. 43, folios 866–67, leg. 1227, PC.

46. José de la Torre to Miguel Cayetano Soler, 5 February 1805, Havana, leg. 2002, SD.

47. Francisco de Arango y Parreño, *Obras del Exmo. Señor D. Francisco Arrango y Parreño*, vol. 1 (Havana: Howson y Heinen, 1888), 461, 466; Guerra y Sánchez et al., *Historia*, 3:9; Moreno Fraginals, *El ingenio*, 57; Thomas, *Cuba*. Despite the prevalence of the story of José de Coca in numerous works, I have found no primary evidence to date in the AGI in Spain, or in the ANC in Cuba, that corroborates this account.

48. Aycock, "Tobacco Regulation," 74–82; L. C. Gray, "The Market Surplus Problems of Colonial Tobacco," *William and Mary Quarterly*, 2nd series, 12, no. 4 (October 1927): 234; Saloutos, "Efforts at Crop Control," 62–65; and Kulikoff, *Tobacco and Slaves*, 109–12.

49. Deans-Smith, *Bureaucrats, Planters, and Workers*, 99.

50. McNeill, *Atlantic Empires*, 198; Marcus Rediker, "'Under the Banner of King Death': The Social World of Anglo-American Pirates, 1716 to 1726," *William and Mary Quarterly*, 3rd series, 38, no. 2 (April 1981): 218. Rediker notes that pirates were prohibited access to clergy, and quotes Vice-Admiralty Judge Nicholas Trott (*The Tryals of Major Stede Bonnet and Other Pirates. . . .* [London, 1719], 2, 3, 4, 34) in his description of them as "'*Hostis Humani Generis*, with whom neither Faith nor Oath' were to be kept, and . . . as 'Brutes, and Beasts of Prey'" (218).

51. Felipe Bellos to Antonio María de Bucarely, 24 August 1766, Sancti Spíritus, doc. 10, leg. 1087, PC; Juan Gutierrez to Bucarely, 31 July 1768, Sancti Spíritus, leg. 1088, PC; Manuel del Portillo to Marqués de la Torre, 13 July 1775, Sancti Spíritus, doc. 79, leg. 1185, PC; del Portillo to de la Torre, 20 July 1775, Sancti Spíritus, doc. 82, leg. 1185, PC.

52. Marqués de la Torre to Antonio Luque, 11 August 1774, Havana, leg. 1144, PC; Marqués de la Torre to Mateo de Echavarría y Elguerra, 13 August 1774, Havana, leg. 1156, PC; Juan Antonio Ayans de Ureta to Pascual de Cisneros andde la Torre, 4 September 1774, Santiago de Cuba, doc. 225, leg. 1142, PC.

53. Marqués de la Torre to José de Albanado, 20 April 1772, Havana, folios 310–11, leg. 1179, PC.
54. Marqués de la Torre to José de Albanado, 10 November 1773, Havana, folios 386–87, leg. 1179, PC; de la Torre to Martín de Asanza, 4 July 1774, Havana, leg. 1179, PC.
55. Marqués de la Torre to Martín de Asanza, 23 September 1774, Havana, leg. 1179, PC; Agustín de Herrera Beloquin to Gobernador General, 1 October 1793, Bayamo, leg. 1464, PC.
56. Juan [*surname unclear*], "Razon de todas las deudas que tiene esta Factoría sublterna de mi cardo con explicacion de anos y demas segun se me pide por el Sr Superintendente director General del ramo," 15 December 1808, Bayamo, exp. 16, leg. 96, IGH.
57. Felipe Bellos to Antonio María de Bucarely, 24 August 1766, Sancti Spíritus, doc. 10, leg. 1087, PC; Marqués de la Torre to Albanado, 20 April 1772, Havana, folios 310–11, leg. 1179, PC; José de Alvaredo to de la Torre, 14 September 1776, Trinidad, leg. 1176, PC; Tomás de la Luz to de la Torre, 25 May 1768, Pinar del Río, leg. 1093, PC; de la Torre to Antonio Fernández, 14 April 1775, leg. 1181, PC; Ramón Lloret to Diego José Navarro, 31 May 1780, Filipinas, leg. 1261, PC; Mier to Navarro, 4 December 1780, Filipinas, leg. 1261, PC; Mier to Navarro, 9 May 1781, Filipinas, leg. 1317, PC; Francisco Cagigal de la Vega to Alonso Arcos y Moreno, 8 May 1752, Havana, order 155, leg. 6, CG.
58. Susan Calafate Boyle, *Los Capitalistas: Hispano Merchants and the Santa Fe Trail* (Alberquerque: University of New Mexico Press, 1997), 56.

CHAPTER 7

1. Fred Anderson describes this period as being about "the control of territory, not thrones." See *Crucible of War: The Seven Years' War and the Fate of Empire in British North America, 1754–1766* (New York: Knopf, 2000), 11.
2. Ramirez de Arellano, "Estado que manifiesta los Tabacos; Josef Pablo Valiente, letter and testimony, 4 July 1799, Havana, no. 72, leg. 2000, SD; Pedro Antonio Gamon to Juan José de la Hoz, 24 January 1804, Real Factoría de Tabacos, Havana, no. 47, leg. 1060, IGH; El Interventor General de la Real Factoría de Tabacos de la Habana Dn. Ignacio Gonzalez Cadura, 3 March 1810, leg. 2000, SD; "Demostración hecha por la Contaduria." By comparison, during the year of greatest tobacco production for the Louisiana and Mexican monopolies, which was also 1788, Louisiana produced in excess of 2 million pounds, or approximately 80,000 *arrobas*, and Mexico's monopoly brought in a little more than 3.5 million pounds, or about 142,000 *arrobas*. Coutts, "Boom and Bust," 306; Deans-Smith, *Bureaucrats, Planters, and Workers*, 263.
3. Sherry Johnson, *Climate and Catastrophe*; Ramirez de Arellano, "Estado que manifiesta los Tabacos." A dip in production following price increases occurred in 1787 for Güines, and in 1801 for all three *partidos* mentioned.
4. Martín Xavier de Echavarría, "Estado en que se manifiesta el peso de 4066 Zurrones de Tavaco. . . ." 24 September 1773, Havana, leg. 1227, PC.
5. Juan Miguel de Arocena, Teniente Gobernador of Cuatro Villas, to Bucarely, 29 October 1768, Trinidad, doc. 46, leg. 1079, PC; Juan Gemir y Lleonart to Antonio María Bucarely, 18 January 1768, Bayamo, leg. 1080, PC.

6. For more on the Hapsburg empire, see M. J. Rodríguez-Salgado, *The Changing Face of Empire: Charles V, Philip II and Habsburg Authority, 1551–1559* (Cambridge: Cambridge University Press, 2008).
7. Bergamini, *Spanish Bourbons*, 22–27. "Near monster and imbecile . . ." appears on page 22.
8. Gonzalo Alvarez, Francisco C. Ceballos, and Celso Quinteiro, "The Role of Inbreeding in the Extinction of a European Royal Dynasty," *PLoS ONE* 4, no. 4 (2009), e5174, *doi:10.1371/journal.pone.0005174*. The study found an inbreeding coefficient (F) of 0.025 for Felipe I, the founder of the Hapsburg dynasty, and a coefficient of 0.254 for Carlos II.
9. Bergamini, *Spanish Bourbons*, 26–29. "Who previously had opened the tombs . . ." appears on page 29.
10. Ibid., 30–31. The French king reportedly told the young man, "Be a good Spaniard; that is now your first duty; but remember that you are born a Frenchman and maintain the unity between the two nations; this is the way to make them happy and to preserve the peace of Europe" (quoted on page 30). An article by Woosang Kim, "Power Transitions and Great War Power from Westphalia to Waterloo," *World Politics* 45, no. 1 (October 1992), lists a figure of 1,251,000 deaths for the twelve-year war (160–61).
11. "Extract of a Letter from Gibraltar, Jan. 17," *Maryland and Baltimore Advertiser*, 14, no. 36: 2; "Extract of a Letter from Plymouth, Dated Sunday Evening, August 22," *Pennsylvania Packet or the General Advertiser* (Philadelphia), 4 November 1779, 2. In his article "Fish for Gold: The Massachusetts Fish Trade with Iberia, 1700–1773," James G. Lydon notes the negative impact of privateering and warships from French Canada on the New England fishing industry (*New England Quarterly* 54, no. 4 [December 1981]: 539–82; see especially 547–48).
12. *The London Chronicle for the Year 1761, from January 1 to June 30*, 9:43, 67. Kris E. Lane's *Pillaging the Empire* (Armonk, NY: M. E. Sharpe, 1998) succinctly describes the fine line between pirate and privateer.
13. Notable exceptions to this trend include Charles H. McCarthy, "The Attitude of Spain during the American Revolution," *Catholic Historical* Review 2, no. 1 (April 1916): 47–65; Leslie Lipson, "European Responses to the American Revolution," in "The American Revolution Abroad," special issue, *Annals of the American Academy of Political and Social Science* 428 (November 1976): 22–32; Andrew J. O'Shaughnessy, "The Formation of a Commercial Lobby: The West India Interest, British Colonial Policy and the American Revolution," *Historical Journal*, 40, no. 1 (March 1997): 71–95; Light Townsend Cummins, *Spanish Observers and the American Revolution, 1775–1783* (Baton Rouge: Louisiana State University Press, 1991); and Thomas E. Chávez, *Spain and the Independence of the United States*.
14. James A. James, "Spanish Influence in the West during the American Revolution," *Mississippi Valley Historical Review* 4, no. 2 (September 1917): 195.
15. Cummins, *Spanish Observers*, 113.
16. Ibid.
17. James A. James, "Spanish Influence in the West," 195, 199–200; Cummins, *Spanish Observers*, 113, 133–36.
18. Nikolaus Böttcher, "Cuba and the Thirteen Colonies during the North American War of Independence," in *Atlantic History: History of the Atlantic System, 1580–1830*,

edited by Horst Pietschmann (Göttingen, Germany: Vandenhoeck and Ruprecht, 2002), 483.

19. D. A. Farnie, "The Commercial Empire of the Atlantic, 1607–1783," *Economic History Review*, 2nd Series, 15, no. 2 (1962): 213.

20. "Extract of a Letter from Bilbao, Dated July 8, to the Honorable Board of War of This State," *Independent Ledger, and the American Advertiser* (Boston) 2, no. 64 (30 August 1779): 3.

21. Francisco Ramirez Portocarro to Don Jose Martinez de Elizalde, 7 November 1780, Madrid, leg. 611, AFTS.

22. "London," *Pennsylvania Journal* (Philadelphia), 9 March 1775, 1.

23. *Pennsylvania Packet or the General Advertiser* (Philadelphia), 23 July 1778, 3.

24. "At a General Meeting of the Freeholders and Inhabitants of the County of Fairfax, on Monday the 18th Day of July 1774," *Boston-Gazette, and Country Journal*, 8 August 1774, 2.

25. *Pennsylvania Ledger* (Philadelphia), 4 February 1775, 2.

26. One such report appeared as "Philadelphia, September 9," *Pennsylvania Packet or the General Advertiser* (Philadelphia), 9 September 1779, 3.

27. "Baltimore, August 9," *Pennsylvania Evening Post* (Philadelphia), 14 August 1780, 95.

28. Unknown to Francisco Florencio de Aguirre, 24 July 1781, Seville, leg. 612, AFTS; Florencio de Aguirre to Josef Martinez de Elizalde, copy of letter, 13 April 1781, Cádiz, leg. 612, AFTS.

29. Francisco Florencio de Aguirre to Manuel Vallerin, copy of letter, 20 October 1780, Cádiz, leg. 611, AFTS; Antonio Josef Bezanes to Josef Martinez de Elizalde, copy of letter, 1 December 1780, Seville, leg. 611, AFTS. "Infamous quality . . ." is from Bezanes.

30. Antonio Garcia Cossio to Marqués de Robledo, 22 October 1779, Cádiz, leg. 610, AFTS; unknown to Administeres Generales de la Renta del Tavaco del Reino, 15 May 1781, Seville, leg. 612, AFTS. For the lack of tobacco from Cuba because of the war, see legs. 611 and 612, AFTS.

31. Don Blas de Hinojosa, certificate, 1744, leg. 250, Ultramar, AGI; "Condiciones de la ultima contracta hecha por Dn Thomas Macones pa provender de Tavacos de Oxa de Virga la Fábricas de Sevilla y Cadiz," n.d., leg. 610, AFTS.

32. O. J. Frederiksen, "Virginia Tobacco in Russia under Peter the Great," *Slavic and East European Review, American Series*, 2, no. 1 (March 1943): 40–44, 55.

33. Barbara W. Tuchman, *The First Salute* (New York: Knopf, 1988), 56; Francisco Florencio de Aguirre to Josef Martinez de Elizalde, 24 March 1781, Cádiz, leg. 612, AFTS.

34. "New-York, January 28," *Independent Gazetteer* (Philadelphia) 1, no. 59 (February 1783): 3.

35. "New-York, May 21," *Salem (MA) Gazette* 30 May 1782, 1, no. 33 (30 May 1782): 3.

36. "Worcester, July 8," *Massachusetts Spy* (Worcester), 9, no. 427 (8 July 1779): 3.

37. "Diary of Grace Growden Galloway, Kept at Philadelphia, July 1, to September 30, 1779," *Pennsylvania Magazine of History and Biography* 58, no. 2 (1934): 152–57, 161–64, 166.

38. "New-York, Sept. 8," *New-York Gazette, and Weekly Mercury* (New York, NY), no. 1456 (13 September 1779): 2.

39. "New-York, September 25," *Royal Gazette* (New York, NY), no. 312 (25 September 1779): 3.
40. "Charlestown (S.C.), October 6," *Virginia Gazette* (Williamsburg), no. 39 (6 November 1779): 2; "Charlestown, October 20," *New-Jersey Gazette* (Burlington) 2, no. 102 (8 December 1779): 2.
41. Antonio Garcia Cossio to Marqués de Robledo, 22 October 1779, Cádiz, leg. 610, AFTS.
42. For more on François-Joseph-Paul, the Count de Grasse (born 1722, died 1788), and the French navy's participation in the American Revolution, see Jonathan Dull, *The French Navy and American Independence: A Study of Arms and Diplomacy, 1774–1787* (Princeton, NJ: Princeton University Press, 1975), and Charles Lee Lewis, *Admiral de Grasse and American Independence* (Annapolis, MD: United States Naval Institute, 1945).
43. "Philadelphia, December 23," *Virginia Gazette* (Williamsburg), no. 50 (22 January 1780): 2.
44. Quoted in Philander D. Chase, "George Washington's Weather," *"Papers of George Washington" News*, no. 4 (Summer 2001): 3, *gwpapers.virginia.edu*.
45. Chase, "George Washington's Weather," 3.
46. Quoted in ibid.
47. William A. Starna and John H. Relethford, "Deer Densities and Population Dynamics: A Cautionary Note," *American Antiquity* 50, no. 4 (October 1985): 827.
48. "Williamsburg, Jan. 8," *Pennsylvania Packet* (Philadelphia), 27 January 1780, 2.
49. "Williamsburg, Jan. 29," *Virginia Gazette* (Williamsburg), no. 51 (29 January 1780): 3.
50. "A Correspondent Has Favored Us with the Following Account of the Weather," *Virginia Gazette* (Williamsburg), no. 49 (15 January 1780): 3.
51. "New-York, Jan. 17," *New-York Gazette; and the Weekly Mercury* (New York, New York), no. 1474 (17 January 1780): 3; "Philadelphia, December 23," *Virginia Gazette* (Williamsburg), no. 50 (22 January 1780): 2; Aileen Moore Topping, "Alexander Guillon in Havana, 'This Very Friendly Port,'" *South Carolina Historical Magazine* 83, no. 1 (January 1982): 46; "New Haven, Jan. 26," *American Journal and General Advertiser* (Providence, RI) 1, no. 48 (10 February 1780): 3; Virginia Department of Emergency Management, "Winter Weather: Virginia Winters," *www.vaemergency.gov/news/history/weather*, accessed 19 October 2012.
52. "New York, January 15," *Royal Gazette*, no. 344 (15 January 1780): 3.
53. "New-York, Jan. 31," *New-York Gazette; and the Weekly Mercury*, no. 1476 (31 January 1780): 3.
54. Untitled article, *New-York Gazette*, no. 1477 (7 February 1780): 3.
55. "New-Haven, February 18," *Pennsylvania Packet, and General Advertiser* (Philadelphia) 13, no. 1668 (4 March 1780): 2.
56. Virginia Department of Emergency Management, "Winter Weather: Virginia Winters."
57. Sherry Johnson, *Climate and Catastrophe*, 144.
58. Ibid., 144–45.
59. Böttcher, "Cuba," 64.
60. Ibid.; James A. Lewis, "Anglo-American Entrepreneurs," 115. For the role of flour shipments from the United States, see also Linda K. Salvucci, "Merchants and Diplomats: Philadelphia's Early Trade with Cuba," *Pennsylvania Legacies* 3, no. 2

(November 2003): 6–10; Linda K. Salvucci, "Atlantic Intersections: Early American Commerce and the Rise of the Spanish West Indies (Cuba)," *Business History Review* 79 (2005): 781–809; and Sherry Johnson, *Climate and Catastrophe*. John Robert McNeill notes that shipments of flour from the United States to Cuba became "more common" after 1748 (*Atlantic Empires*, 177). For *tasajo*, see José Luis Luzon, "Comer y beber en La Habana colonial," *Caravelle* 71 (December 1998): 30–31; and Claudio Saunt, *West of the Revolution: An Uncommon History of 1776* (New York: Norton, 2014).

61. Quoted in *New-Haven (CT) Gazette* 1, no. 8 (1 July 1784): 1.
62. *Dunlap's American Daily Advertiser* (Philadelphia), no. 4475 (29 May 1793): 3.
63. José de Ymaz et al., "Condiciones con que se ha verificado."
64. Luis de Viguri to Miguel Cayetano Soler, 14 March 1800, Havana, leg. 2000, SD; unknown to Julián de Arriaga, 8 October 1771, Havana, doc. 70, leg. 1140, PC. For the many wars that beset Atlantic shipping during this period, see Noel Mostert, *The Line upon a Wind: The Great War at Sea, 1793–1815* (New York: Norton, 2007).
65. R. H. Wyman, comp., *Winds, Currents, and Navigation of the Gulf of Cadiz, the Western Coast of the Spanish Peninsula, and the Strait of Gibraltar, Compiled from the Most Reliable Authorities* (Washington, DC: Government Printing Office, 1870), 3–4; Robert F. Marx, *Shipwrecks in the Americas* (Mineola, NY: Dover, 1987 [1971]), 5; J. H. Elliott, *Empires of the Atlantic World: Britain and Spain in America, 1492–1830* (New Haven, CT: Yale University Press, 2006), 226. "Bring heavy . . ." is from Wyman, *Winds, Currents, and Navigation*, 3.
66. Leg. 685, AFTS (this entire legajo is dedicated to tobacco coming from Havana in 1789).
67. Warren, *Vagamundo*, 150–52; Fermín Arana de Varaflora, *Compendio historico descriptivo de la muy noble y muy leal ciudad de Sevilla metropolí de Andalucía formabalo don Fermín Arana de Varaflora, natural y vecino de dicha Ciudad* (1789), 82–83; José M. Rodríguez Gordillo, *Un archivo para la historia del Tabaco* (Madrid: Jacaryan, 1984), 23–49; José M. Rodríguez Gordillo, José Morales Sánchez, and Alfonso E. Pérez Sánchez, *Goya: Retratos para la Real Fábrica de Tabacos de Sevilla* (Seville: Edita Tabapress para Tabacalera, 1985), 14–15. Arana de Varaflora noted that twenty-one of the machines were powered by six *molinos* each, while there were six machines with four mills each, two with two mills each, and forty-eight additional mills.

EPILOGUE

1. Fernando VII, "El Rey: Dedicado desde la divina Providencia me ha resituido al trono de mis amados vasallos, formentando la agricultura, el comercio y la industria de mis reinos. . . ." 23 June 1817, Madrid, leg. 2000, SD; Rivero Muñiz, *Tabaco*, 2:231–32.
2. Agustín Hernández Pulgarón, *tazmia*, Partidos de Bacuranado y San Jeronimo, 6 April 1770, Guanabacoa, leg. 1090, PC; Luis Echavarría, *tazmia*, Partido de Mayarí, 12 April 1770, Mayarí, leg. 1080, PC; Joaquín de Sola, *tazmia*, Partidos de Concepción al Norte, Saramaguacan al Norte, Arenillas al Norte, Guariao al Sur, and Guanausi al Sur, 25 May 1770, Puerto Príncipe, leg. 1085, PC; Ramón José Rodríguez, *tazmia*, Partido de Yumuri, 26 May 1770, Matanzas, leg. 1090, PC; Ramón José Rodríguez, *tazmia*, Partidos de Naranjal, Estancias y San Agustín, 15 May 1770, leg. 1090, PC; Guerra y Sánchez et al., *Historia*, 3:9; José González, "Re-

sumen general de la visita. . . ." 31 May 1817, Real Factoría de Tabacos, Havana, leg. 2019, SD.

3. Sherry Johnson, *Social Transformation*, 172–73.
4. "Documentos para la historia económica y social de Cuba," *Boletín del Archivo Nacional de Cuba*, 53–54 (1953–1954): 280–83, quoted in Sherry Johnson, *Social Transformation*, 173.
5. Ortiz, *Cuban Counterpoint*, 59; Sherry Johnson, *Social Transformation*, 56, 172–73. "The official guardian of their interests" is from Johnson, 173. Preferential treatment for *vegueros* was not just limited to Cuba. In Mexico, sugar *hacendados* complained that the Crown unfairly intervened on behalf of tobacco farmers. See Deans-Smith, *Bureaucrats, Planters, and Workers*, 108.
6. Ramires de Arellano, "Estado que manifiesta los Tabacos."
7. De la Sagra, *Historia economico-politica y estadistica*, 240–66, cited in Carlos Marichal, *Bankruptcy of Empire: Mexican Silver and the Wars between Spain, Britain, and France 1760–1810* (Cambridge: Cambridge University Press, 2007), 38. As noted in an earlier chapter, scholars argue that a sizable amount of tobacco was grown and traded clandestinely and never made it to the official market.
8. Urriza, "Nomina"; don Braulio de Vivanco, employment record, circa 31 August 1817, Havana, leg. 2019, SD.
9. Rivero Muñiz, *Las tres sediciones*, 20; Marichal, *Bankruptcy of Empire*, 38.
10. "Noticia del estado de posesiones."
11. Ramón Lloret, *tazmia*, 24 April 1780, Filipinas, exp. 156, folio 228v, leg. 1261, PC; Rebecca J. Scott, *Slave Emancipation in Cuba*, 263; Guerra y Sánchez, *Sugar and Society*, 49; Guerra y Sánchez et al., *Historia*, 3:178; Jacobo de la Pezuela, "Estado general de la produccion agricola y riqueza rural de la isla correspondiente al año de 1861," in *Diccionario geográfico, estadistico, histórico de la Isla de Cuba*, vol. 1 (Madrid: Imprenta del Establecimiento de Mallado, 1863), 38–39; Jean Stubbs, *Tabaco en la periferia: El complejo agro-industrial cubano y su movimiento obrero, 1860–1959* (Havana: Editorial de Ciencias Sociales, 1989), 211. Stubbs notes that even as late as 1862, the vast majority of slaves working in tobacco (12,174 out of 17,675) were found in Pinar del Río (222), while Louis A. Pérez Jr. argues that most slaves, 172,671 or 46.9 percent of Cuba's total 1862 slave population, were found on *ingenios* (*Winds of Change*, 105).
12. Pérez, *Winds of Change*, 69–70, 79–80, 94.
13. Marrero, *Cuba*, 11:60–61; Pérez, *Winds of Change*, 95; Nuñez and González, "Plan demostrativo"; Stubbs, *Tabaco en la periferia*, 211; US War Department, Office Director Census of Cuba, *Report on the Census of Cuba, 1899* (Washington, DC: Government Printing Office, 1900), 549–50.
14. Pérez, *Cuba between Empires*, 4–7; Ferrer, *Insurgent Cuba*, 6–42, 72; Vanessa Michelle Ziegler, "The Revolt of the 'Ever-Faithful Isle': The Ten Years' War in Cuba, 1868–1878" (PhD diss., University of California, Santa Barbara, 2007), 254–57.
15. C. Neale Ronning, *José Martí and the Émigré Colony in Key West: Leadership and State Formation* (New York: Praeger, 1990); Gerald E. Poyo, "Cuban Patriots in Key West, 1878–1886: Guardians at the Separatist Ideal," *Florida Historical Quarterly* 61, no. 1 (July 1982): 20–36; Pérez, *Cuba between Empires*, xv–xx; Louis A. Pérez Jr., *Cuba and the United States: Ties of Singular Intimacy*, 3rd ed. (Athens: University of Georgia Press, 2003), 87–88.

16. John Lawrence Tone, *War and Genocide in Cuba, 1895–1898* (Chapel Hill: University of North Carolina Press, 2006), 223; Pérez, *Winds of Change*, 47–48, 54.
17. Louis A. Pérez Jr., *Lords of the Mountain: Social Banditry and Peasant Protest in Cuba, 1878–1918* (Pittsburgh: University of Pittsburgh Press, 1989), 61.
18. Robert Percival Porter, *Appendix to the Report on the Commercial and Industrial Condition of the Island of Cuba* (Washington, DC: Government Printing Office, 1899), 208.
19. Other products were similarly controlled by the Bourbons through royal monopolies, like the popular cane brandy known as *aguardiente*. See Anthony McFarlane, *Colombia before Independence: Economy, Society, and Politics under Bourbon Rule*, Cambridge Latin American Studies (Cambridge: Cambridge University Press, 1993), 200–201; and Gilma Lucía Mora de Tovar, *Aguardiente y conflictos sociales en la Nueva Granada durante el siglo XVIII* (Bogotá: Universidad Nacional de Colombia, Centro Editorial, 1988).
20. Kuethe and Inglis, "Absolutism," 122.
21. Ibid., 121–22. Along with Ricardo Wall, who held the dual positions of Spain's minister of state and war, Arriaga took responsibility for the loss of the Cuban capital to the British. Arriaga's position at court weakened as a result.
22. Julián de Arriaga to Antonio María de Bucarely, 18 July 1769, Madrid, leg. 1134, PC; Arriaga to Bucarely, 22 October 1769, San Lorenzo, folio 54–56, leg. 1134, PC; Arriaga to Bucarely, 22 July 1770, Aranjuez, folio 100, leg. 1134, PC; Arriaga to Bucarely, 22 August 1769, Madrid, folio 34, leg. 1134, PC; Arriaga to Bucarely, 22 September 1769, San Ildefonso, folio 56, leg. 1134, PC; Arriaga to Bucarely, 20 July 1769, Madrid, folio 28, leg. 1134, PC; Arriaga to Bucarely, 22 May 1770, folio 92, leg. 1134, PC; Arriaga to Bucarely, San Ildefonso, 22 August 1769, folio 50, leg. 1134, PC; Arriaga to Bucarely, 22 February 1770, Madrid, folios 68–69, leg. 1134, PC; Arriaga to Bucarely, 18 July 1769, Madrid, folio 24, leg. 1134, PC.
23. Kuethe, *Cuba*, 77; Miguel de Múzquiz to Marqués de la Torre, 21 December 1775, Madrid, exp. 52, folios 884–86, leg. 1227, PC; Luis de Viguri to Miguel Cayetano Soler, 28 June 1800, Havana, leg. 2000, SD.
24. Julián de Arriaga to Antonio María de Bucarely, 22 September 1769, San Ildefonso, folio 46, leg. 1134, PC; Arriaga to Bucarely, 22 May 1770, Aranjuez, folio 92, leg. 1134, PC; Arriaga to Bucarely, 24 May 1771, Aranjuez, leg. 1140, PC.
25. *Cigar Aficionado*, "The Twenty-Five Best Cigars of the Year," *www.cigaraficionado.com/top25*, accessed 4 April 2012. For the US prohibition on importing Cuban tobacco products, including cigars, see US Customs and Border Protection, "Importing Cuban Cigars," *help.cbp.gov*, accessed 19 June 2014.
26. Borsay, *History of Leisure*, 169; Todd Butler, "Power in Smoke," 118. Inter-cultural exchanges between Cuba and the United States are found in Louis A. Pérez Jr., *On Becoming Cuban: Identity, Nationality, and Culture* (Chapel Hill: University of North Carolina Press, 1999).
27. Pierre Salinger, "Kennedy, Cuba, and Cigars," *Cigar Aficionado* 1, no. 1 (Autumn 1992), *www.cigaraficionado.com*; Pierre Salinger, *P.S.: A Memoir* (New York: Macmillan, 2001), 239–49. In his 2001 memoir, Salinger describes the episode as "one of my favorite cigar stories of all time."
28. Salinger, "Kennedy, Cuba, and Cigars."

Bibliography

Abbot, Abiel. *Letters Written in the Interior of Cuba, between the Mountains of Arcana, to the East, and of Cusco, to the West, in the Months of February, March, April, and May, 1828*. Boston: Bowles and Dearborn, 1829.

Aimes, Hubert H. S. *A History of Slavery in Cuba, 1511–1868*. New York: Octagon, 1967.

Alford, B. W. E. *W. D. and H. O. Wills and the Development of the U.K. Tobacco Industry, 1786–1965*. London: Methuen, 1973.

Alvarez, Gonzalo, Francisco C. Ceballos, and Celso Quinteiro. "The Role of Inbreeding in the Extinction of a European Royal Dynasty." *PLoS ONE* 4, no. 4 (2009): e5174. doi:10.1371/journal.pone.0005174.

Un Amigo del Pais. *Memoria sobre el tabaco de la isla de Cuba, en la que se indican algunas mejoras de que es susceptible su cultivo y preparacion*. Havana: Imprenta de Antonio María Davila, 1852.

Ancheta Niebla, Eduarda. "La Orden de San Juan de Dios y el Hospital de San Felipe y Santiago." *Enfermería Avanza*, 19 April 2012, *enfeps.blogspot .com.es*. PDF version accessed at Nicanor Aniorte Hernández, *Página personal sobre la Formación Postgrado en Enfermería*, *www.aniorte-nic.net*.

Anderson, Fred. *Crucible of War: The Seven Years' War and the Fate of Empire in British North America, 1754–1766*. New York: Knopf, 2000.

Anderson, Susan Campbell. "A Matter of Authority: James I and the Tobacco War." *Comitatus: A Journal of Medieval and Renaissance Studies* 29, no. 1 (1998): 136–63.

Andrews, Charles M., ed. *Narratives of the Insurrections, 1675–1690*. 1915. New York: Barnes and Noble, 1967.

Andrews, Kenneth R. *Elizabethan Privateering: English Privateering during the Spanish War, 1585–1603*. Cambridge: Cambridge University Press, 1964.

———. *The Spanish Caribbean: Trade and Plunder, 1530–1630*. New Haven, CT: Yale University Press, 1978.

Arana de Varaflora, Fermín. *Compendio historico descriptivo de la muy noble y muy leal ciudad de Sevilla metropolí de Andalucía formabalo don Fermín Arana de Varaflora, natural y vecino de dicha Ciudad*, 1789.

Arango y Parreño, Francisco de. *Obras del Exmo. Señor D. Francisco Arrango y Parreño*. Vol. 1. Havana: Howson y Heinen, 1888.

Araúz Monfante, Celestino Andrés. "La acción ilegal de los holandeses en el Caribe y su impacto en las Antillas y Puerto Rico durante la primera mitad del siglo XVIII." *Revista/Review Interamericana* 14, nos. 1–4 (Spring/Winter 1984): 67–79.

Arcila Farias, Eduardo. *Historia de un monopolio: El estanco del tabaco en Venezuela 1779–1833*. Caracas: Ediciones de la Facultad de Humanidades y Educación, Universidad Central de Venezuela, 1977.

Armíldez de Toledo, Conde. *Noticias estadisticas de la Isla de Cuba en 1862, dispuestas y publicadas por el centro de estadistica*. Havana: Imprenta del Gobierno, Capitania General y Real Hacienda por S. M., 1864.

Armitage, David. "Three Concepts of Atlantic History." In *The British Atlantic World, 1500–1800*, edited by David Armitage and Michael J. Braddick, 11–27. New York: Palgrave Macmillan, 2002.

Armytage, Frances. *The Free Port System in the British West Indies: A Study in Commercial Policy, 1766–1822*. London: Longmans, Green & Company, 1953.

Axton, W. F. *Tobacco and Kentucky*. Lexington: University Press of Kentucky, 1975.

Aycock, William Brantley. "Tobacco Regulation in Colonial Virginia." MA thesis, University of North Carolina, 1937.

Ballou, Maturin M. *Due South; or, Cuba, Past and Present*. Boston: Houghton, Mifflin, 1885.

Barbier, Jacques A., and Herbert S. Klein. "Revolutionary Wars and Public Finances: The Madrid Treasury, 1784–1807." *Journal of Economic History* 41, no. 2 (June 1981): 315–39.

Barbier, Jacques A., and Allan J. Kuethe, eds. *The North American Role in the Spanish Imperial Economy, 1760–1819*. Manchester, UK: Manchester University Press, 1984.

Barbour, Violet. "Privateers and Pirates of the West Indies," *American Historical Review* 16, no. 3 (April 1911): 529–66.

Barickman, B. J. "A Bit of Land, Which They Call *Roca*: Slave Provision Grounds in the Bahian Reconcavo, 1780–1860." *Hispanic American Historical Review* 74, no. 4 (November 1994): 649–87.

Baud, Michiel. *Peasants and Tobacco in the Dominican Republic, 1870–1930*. Knoxville: University of Tennessee Press, 1995.

Bergamini, John D. *The Spanish Bourbons: The History of a Tenacious Dynasty*. New York: Putnam, 1974.

Bielenberg, Andy, and David Johnson. "The Production and Consumption of Tobacco in Ireland, 1800–1914." *Irish Economic and Social History* 25 (1998): 1–21.

Bitterli, Urs. *Cultures in Conflict: Encounters between European and Non-European Cultures, 1492–1800*. Stanford, CA: Stanford University Press, 1989.

Blanton, Dennis B., Michael Chenoweth, and Cary J. Mock. "The Great Flood of 1771: An Explanation of Natural Causes and Social Effects." In Dupigny-Giroux and Mock, *Historical Climate Variability*, 3–21.

Boone, M. Elizabeth. *Vistas de España: American Views of Art and Life in Spain, 1860–1914*. New Haven, CT: Yale University Press, 2007.

Borsay, Peter. *A History of Leisure: The British Experience since 1500*. New York: Palgrave Macmillan, 2006.

Boswell, Thomas D. "The West Indies: The Hispanic Territories and Haiti." In *Middle America: Its Lands and Peoples*, edited by Robert C. West, John P. Augelli, Thomas D. Boswell et al., 128–67. Englewood Cliffs, NJ: Prentice Hall, 1989.

Böttcher, Nikolaus. "Cuba and the Thirteen Colonies during the North American War of Independence." In *Atlantic History: History of the Atlantic System, 1580–1830*, edited by Horst Pietschmann, 481–95. Göttingen, Germany: Vandenhoeck and Ruprecht, 2002.

Bowser, Frederick P. "The Free Person of Color in Mexico City and Lima: Manumission and Opportunity, 1580–1650." In *Race and Slavery in the Western Hemisphere: Quantitative Studies*, edited by Stanley L. Engerman and Eugene D. Genovese, 331–68. Princeton, NJ: Princeton University Press, 1975.

Boyle, Susan Calafate. *Los Capitalistas: Hispano Merchants and the Santa Fe Trail*. Albuquerque: University of New Mexico Press, 1997.

Breen, T. H. *Tobacco Culture: The Mentality of the Great Tidewater Planters on the Eve of the Revolution*. 2nd ed. Princeton, NJ: Princeton University Press, 2001.

Brennan, W. A. *Tobacco Leaves: Being a Book of Facts for Smokers*. Menasha, WI: Collegiate Press, 1915.

Brown, Kendall W. "Jesuit Wealth and Economic Activity with the Peruvian Economy: The Case of Colonial Southern Peru." *Americas* 44, no. 1 (July 1987): 23–43.

Brown, Vera Lee. "The South Sea Company and Contraband Trade." *American Historical Review* 31, no. 4 (July 1926): 662–78.

Burke, Peter. "Viewpoint: The Invention of Leisure in Early Modern Europe." In *Past and Present* 146 (February 1995): 136–50.

Butler, Todd. "Power in Smoke: The Language of Tobacco and Authority in Caroline England." *Studies in Philology* 106, no. 1 (Winter 2009): 100–118.

Carlson, Fred A. *Geography of Latin America*. 3rd ed. New York: Prentice Hall, 1952.

Carmagnani, Marcello. "La oposición a los tributos en la segunda mitad del siglo XVIII." *Revista chilena de historia y geografía* 129 (1961): 158–95.

Chase, Philander D. "George Washington's Weather." "*Papers of George Washington" News* 4 (Summer 2001), *gwpapers.virginia.edu*.

Charlton, Anne. "Medicinal Uses of Tobacco in History." *Journal of the Royal Society of London* 97 (June 2004): 292–96.

———. "Tobacco or Health 1602: An Elizabethan Doctor Speaks." *Health Education Research* 20, no. 1 (2005): 101–11.

Chávez, Thomas E. *Spain and the Independence of the United States: An Intrinsic Gift*. Albuquerque: University of New Mexico Press, 2002.

Cluster, Dick, and Rafael Hernández. *The History of Havana*. New York: Palgrave Macmillan, 2006.

Coatsworth, John H. "The Mexican Mining Industry in the Eighteenth Century." In Jacobsen and Puhle, *Economies*, 26–45.

Cohen, Jeremy. "Smuggling in Eighteenth-Century Venezuela: The Role of the *Práctico*." Paper presented at the meeting of the Latin American Studies Association, Dallas, March 2003.

Columbus, Christopher. *The "Diario" of Christopher Columbus's First Voyage to America, 1492–1493*. Abstracted by Fray Bartolomé de las Casas. Transcribed and translated into English, with notes and concordance of the Spanish, by Oliver Dunn and James E. Kelley Jr. Norman: University of Oklahoma Press, 1989.

———. *The Journal of Christopher Columbus (during his First Voyage, 1492–93), and Documents Relating to the Voyages of John Cabot and Gaspar Corte Real*. Translated with notes and an introduction by Clements R. Markham. London: Hakluyt Society, 1893.

Corbitt, Duvon Clough. "The Colonial Government of Cuba." PhD diss., University of North Carolina, 1938.

Courtwright, David. *Forces of Habit: Drugs and the Making of the Modern World*. Cambridge, MA: Harvard University Press, 2001.

Coutts, Brian E. "Boom and Bust: The Rise and Fall of the Tobacco Industry in Spanish Louisiana, 1770–1790." *Americas* 42, no. 3 (January 1986): 289–309.

Cruz, Asiris M. "Smoking Guns: Tobacco Farmers' Revolts and Colonial Society, Cuba, 1650–1750." PhD diss., University of California, Riverside, 2007.

Cuadrado Melo, Manuel. *Obispado de la Habana su historia a traves de los siglos*. Vol. 1, part 1A. Havana: n.p., 1970.

Cummins, Light Townsend. *Spanish Observers and the American Revolution, 1775–1783*. Baton Rouge: Louisiana State University Press, 1991.

Cushner, Nicholas P. *Lords of the Land: Sugar, Wine, and Jesuit Estates of Coastal Peru, 1600–1767*. Albany: State University of New York Press, 1980.

Dabney, Virginius. *Virginia: The New Dominion*. 1971. Charlottesville: University Press of Virginia, 1996.

Dana, Richard Henry, Jr. *To Cuba and Back*. Edited with an introduction by C. Harvey Gardiner. 1859. Carbondale: Southern Illinois University Press, 1966.

Davies, K. G. *The North Atlantic World in the Seventeenth Century*. Minneapolis: University of Minnesota Press, 1974.

Davillier, Charles. *Spain*. Translated by J. Thomson. 1874. New York: Scribner, Welford, and Armstrong, 1876.

Deans-Smith, Susan. *Bureaucrats, Planters, and Workers: The Making of the Tobacco Monopoly in Bourbon Mexico*. Austin: University of Texas Press, 1992.

———. "The Money Plant: The Royal Tobacco Monopoly of New Spain, 1765–1821." In Jacobsen and Puhle, *Economies*, 361–87.

de Jesus, Ed. C. *The Tobacco Monopoly in the Philippines: Bureaucratic Enterprise and Social Change, 1760–1880*. Manila: Ateneo de Manila University Press, 1980.

de la Pezuela, Jacobo. "Estado general de la produccion agricola y riqueza rural de la isla correspondiente al año de 1861." In *Diccionario geográfico, estadistico, histórico de la Isla de Cuba*, 1:38–39. Madrid: Imprenta del Establecimiento de Mallado, 1863.

———. "Tabaco." In *Diccionario geográfico, estadistico, histórico de la Isla de Cuba*, 4:566. Madrid: Imprenta del Establecimiento de Mallado, 1867.

de la Sagra, Ramón. *Historia economico-politica y estadistica de la Isla de Cuba ó sea de sus progresos en la población, la agricultura, el comercio y las rentas*. Havana: Imprenta de las Viudas de Arazoza y Soler, 1831.

———. *Historia fisica, politica y natural de la isla de Cuba*. Paris: A. Bertrand, 1839.

de Paula Arias, Antonio María. *El veguero de Vuelta Abajo: Apuntes sobre el cultivo del tabaco*. Pinar del Río, Cuba: Imprenta de Miguel Vives, 1887.

de Paula Sanz, Francisco. *Viaje por el Virreinato del Río de la Plata: El camino del tabaco*.

Buenos Aires: Centro de Estudios Interdisciplinarios de Hispanoamérica Colonial, 1977.

Díaz, Bernal. *The Conquest of New Spain*. Translated and with an introduction by J. M. Cohen. 1963. New York: Penguin, 1988.

Dimock, Joseph Judson. *Impressions of Cuba in the Nineteenth Century: The Travel Diary of Joseph J. Dimock*. Edited by Louis A. Pérez Jr. Wilmington, DE: Scholarly Resources, 1998.

Dollero, Adolfo. *Cultura cubana: La provincia de Pinar del Río y su evolucíon*. Havana: Seoane, Fernández, 1921.

Drelichman, Mauricio, and Hans-Joachim Voth. "The Sustained Debts of Philip II: A Reconstruction of Castile's Fiscal Position, 1566–1596," *The Journal of Economic History* 70, no. 4 (December 2010): 813–42.

Dull, Jonathan. *The French Navy and American Independence: A Study of Arms and Diplomacy, 1774–1787*. Princeton, NJ: Princeton University Press, 1975.

Dupigny-Giroux, Lesley-Ann, and Cary J. Mock, eds. *Historical Climate Variability and Impacts in North America*. Dordrecht, Netherlands: Springer, 2009.

Elliott, J. H. *Empires of the Atlantic World: Britain and Spain in America, 1492–1830*. New Haven, CT: Yale University Press, 2006.

Elsner, Eleanor. *Romantic Spain: The Land of Mantillas and Matadors*. London: Butterworth, 1924.

Ennis, Huberto M. "Shortages of Small Change in Early Argentina." WP 03-12, Working Paper Series. Richmond, VA: Federal Reserve Bank of Richmond, 2003.

"Establiciendo una Compañía en San Cristóbal de la Habana Dedicada a la Conducción de Tabacos, Azúcar, Corambre y Otros Frutos de la Isla de Cuba." In *Cedulario Americano del Siglo XVIII*, vol. 3, *Cédulas de Luis I (1724), Cédulas de Felipe V (1724–1746)*, edited by Antonio Muro Orejon, 253–76. Seville: Escuela de Estudios Hispano-Americanos, 1740.

Fallas, Marco Antonio. *La factoría de tabacos de Costa Rica*. San Jose: Editorial Costa Rica, 1972.

Farnie, D. A. "The Commercial Empire of the Atlantic, 1607–1783." *Economic History Review*, 2nd series, 15, no. 2 (1962): 205–18.

Felipe V. "Instrucción dada por Felipe V el día 11 de abril de 1717 para el estanco de tabaco." In *Documentos para la historia de Cuba*, vol. 1, edited by Hortensia Pichardo, 147–48. Havana: Editorial de Ciencias Sociales, 1971.

Ferrer, Ada. *Insurgent Cuba: Race, Nation, and Revolution, 1868–1898*. Chapel Hill: University of North Carolina Press, 1999.

Firestone, Harry, ed. *Bacon's Rebellion: The Contemporary News Sheets*. Charlottesville: University of Virginia Press, 1956.

Foner, Philip Sheldon. *Antonio Maceo: The "Bronze Titan" of Cuba's Struggle for Independence*. New York: Monthly Review Press, 1977.

Frederiksen, O. J. "Virginia Tobacco in Russia under Peter the Great." In *Slavic and East European Review, American Series* 2, no. 1 (March 1943): 40–56.

Friederici, Georg. *Der Charakter der Entdeckung und Eroberung Amerikas durch die Europäer*. Vol. 1. Stuttgart: Perthes, 1925.

Friedländer, H. E. *Historia económica de Cuba*. Vol. 14. Havana: Jesus Montero, 1944.

García Galló, Gaspar Jorge. *Biografía del tabaco habano*. Havana: Universidad Central de Las Villas, Departamento de Relaciones Culturales, 1959.

García Rodríguez, Mercedes. "Presencia Jesuita en la Economía de Cuba: Siglo XVIII." PhD diss., Universidad de la Habana, 1999.

González Enciso, Agustín. "Los usos financieros del tabaco; o bien, para qué servían los dineros obtenidos con la renta del tabaco." In *El mercado del tabaco en España durante el siglo XVIII: Fiscalidad y consumo*, edited by Santiago de Luxán Meléndez, Sergio Solbes Ferri, and Juan José Laforet, 25–51. Las Palmas, Spain: Universidad de Las Palmas de Gran Canaria Servicio de Publicaciones, 2000.

González-Ripoll, María Dolores, and Izaskun Álvarez Cuartero, eds. *Francisco Arango y la invención de la Cuba azucarera*. Salamanca, Spain: Ediciones Universidad Salamanca, 2009.

Goodman, Jordan. *Tobacco in History: The Cultures of Dependence*. London: Routledge, 1993.

Grahn, Lance. *The Political Economy of Smuggling: Regional Informal Economies in Early Bourbon New Granada*. Dellplain Latin America Studies, no. 35. Edited by David R. Robinson. Boulder, CO: Westview Press, 1997.

Gray, L. C. "The Market Surplus Problems of Colonial Tobacco." *William and Mary Quarterly* 2nd series, 12, no. 4 (October 1927): 231–45.

Gray, Stanley, and V. J. Wyckoff. "The International Tobacco Trade in the Seventeenth Century." *Southern Economic Journal* 7, no. 1 (July 1940): 1–26.

Guerra y Sánchez, Ramiro. *Sugar and Society in the Caribbean: An Economic History of Cuban Agriculture*. With a foreword by Sidney W. Mintz. Caribbean Series. New Haven, CT: Yale University Press, 1964.

Guerra y Sánchez, Ramiro, José M. Pérez Cabrera, Juan J. Remos, and Emeterio S. Santovenia. *Historia de la nación cubana*. Vol. 2, *Guerras coloniales conflictos y progresos (desde 1697 hasta 1790)*. Havana: Editorial Historia de la Nacion Cubana, 1952.

———. *Historia de la nación cubana*. Vol. 3, *Ilustración libertad de comercio (desde 1790 hasta 1837)*. Havana: Editorial Historia de la Nacion Cubana, 1952.

Gutiérrez, Fernando, Carlos Larrinaga, and Miriam Núñez. "Cost and Management Accounting in Pre-industrial Revolution Spain." *Accounting Historians Journal* 32, no. 1 (June 2005), 111–48.

Gutiérrez Escudero, Antonio. "El tabaco en Santo Domingo y su exportación a Sevilla (época colonial)." In *Relaciones de poder y comercio colonial: Nuevas perspectivas*, edited by Enriqueta Vila Vilar and Allan J. Kuethe, 117–42. Seville: Escuela de Estudios Hispano-Americanos de Sevilla, 1999.

Gutman, Herbert G. *The Black Family in Slavery and Freedom, 1750–1925*. New York: Vintage, 1976.

Hahn, Barbara. *Making Tobacco Bright: Creating an American Commodity, 1617–1937*. Baltimore: Johns Hopkins University Press, 2011.

Hall, Gwendolyn Midlo. *Slavery and African Ethnicities in the Americas: Restoring the Links*. Chapel Hill: University of North Carolina Press, 2005.

Hanson, Carl A. "Monopoly and Contraband in the Portuguese Tobacco Trade, 1624–1702." *Luso-Brazilian Review* 19, no. 2 (Winter 1982): 149–68.

Harrison, John P. "The Evolution of the Colombian Tobacco Trade, to 1875." *Hispanic American Historical Review* 32, no. 2 (May 1952): 163–74.

Hazard, Samuel. *Cuba with Pen and Pencil*. 1871. Miami: Editorial Cubana, 1989.

Helg, Aline. *Our Rightful Share: The Afro-Cuban Struggle for Equality, 1886–1912*. Chapel Hill: University of North Carolina Press, 1995.

Hernández García, Julio. *La emigración canaria contemporánea (Siglo XIX)*. Las Palmas, Spain: Ediciones del Cabildo Insular de Gran Canaria, 1987.

Hernández González, Manuel. *Canarias: La emigración*. La Laguna, Tenerife: Impecan, 1995.

Herndon, G. Melvin. *William Tatham and the Culture of Tobacco*. Coral Gables, FL: University of Miami Press, 1969.

Hilton, Matthew. "Leisure, Politics, and the Consumption of Tobacco in Britain since the Nineteenth Century." In *Histories of Leisure*, edited by Rudy Koshar, 319–36. Oxford: Berg, 2002.

Humboldt, Alexander von. *The Island of Cuba*. Translated by J. S. Thrasher. New York: Derby and Jackson, 1856.

[Hurlbert, William Henry]. *Gan-Eden; or, Pictures of Cuba*. Boston: John P. Jewett, 1854.

James, James A. "Spanish Influence in the West during the American Revolution." *Mississippi Valley Historical Review* 4, no. 2 (September 1917): 193–208.

James I. *A Counterblaste to Tobacco*. London: 1604.

Johnson, Paul R. *The Economics of the Tobacco Industry*. New York: Praeger, 1984.

Johnson, Sherry. *Climate and Catastrophe in Cuba and the Atlantic World in the Age of Revolution*. Chapel Hill: University of North Carolina Press, 2011.

———. "From Authority to Impotence: Arango's Adversaries and Their Fall from Power during the Constitutional Period (1808–1823)." In González-Ripoll and Álvarez Cuartero, *Francisco Arango*, 193–211.

———. "'*La Guerra Contra los Habitantes de los Arrabales*': Changing Patterns of Land Use and Land Tenancy in and Around Havana, 1763–1800." *Hispanic American Historical Review* 77, no. 2 (1997): 181–209.

———. "'*Señoras en Sus Clases no Ordinarias*': Enemy Collaborators or Courageous Defenders of the Family?" *Cuban Studies/Estudios Cubanos* 34 (2004): 25–30.

———. *The Social Transformation of Eighteenth-Century Cuba*. Gainesville: University Press of Florida, 2001.

———. "Where Has All the Flour Gone? El Niño, Environmental Crisis and the Emergence of Alternative Markets in the Hispanic Caribbean, 1760s–1770s." *William and Mary Quarterly* 62, no. 3 (July 2005): 365–410.

Jones, J. A. P. *Europe, 1500–1600*. Surrey, UK: Thomas Nelson, 1997.

Karasch, Mary. "Slave Women on the Brazilian Frontier in the Nineteenth Century." In *More Than Chattel: Black Women and Slavery in the Americas*, edited by David Barry Gaspar and Darlene Clark Hine, 79–96. Bloomington: Indiana University Press, 1996.

Kim, Woosang. "Power Transitions and Great War Power from Westphalia to Waterloo." In *World Politics* 45, no. 1 (October 1992): 153–72.

Kiple, Kenneth F. *Blacks in Colonial Cuba, 1774–1899*. Latin American Monographs, 2nd Series. Gainesville: University Presses of Florida, 1976.

Kipling, Rudyard. *The Collected Poems of Rudyard Kipling*. Ware, UK: Wordsworth Editions, 1994.

Klein, Herbert S. *Slavery in the Americas: A Comparative Study of Virginia and Cuba*. Chicago: University of Chicago Press, 1967.

Klooster, Willem Wubbo. "Illicit Riches: The Dutch Trade in the Caribbean, 1648–1795." PhD diss., Rijksuniversiteit te Leiden (Netherlands), 1995.

Knight, Franklin W. *Slave Society in Cuba during the Nineteenth Century*. Madison: University of Wisconsin Press, 1970.

Kuethe, Allan J. *Cuba, 1753–1815: Crown, Military, and Society.* Knoxville: University of Tennessee Press, 1986.

———. "The Development of the Cuban Military as a Sociopolitical Elite, 1763–83." *Hispanic American Historical Review* 61, no. 4 (1981): 695–704.

———. "El situado mexicano, los azucareros y la fielded cubana," in *Las Antillas en la era de las luces y la revolución*, edited by José Antonio Piqueras Arenas, (Madrid: Siglo XXI de España Editores, 2005), 301–18.

———. "*Los Llorones Cubanos*: The Socio-military Basis of Commercial Privilege in the American Trade under Charles IV." In Barbier and Kuethe, *North American Role*, 142–56.

Kuethe, Allan J., and G. Douglas Inglis. "Absolutism and the Enlightened Reform: Charles III, the Establishment of the *Alcabala*, and Commercial Reorganization in Cuba." *Past and Present: A Journal of Historical Studies* 109 (November 1985): 118–43.

Kulikoff, Alan. *Tobacco and Slaves: The Development of Southern Cultures in the Chesapeake, 1680–1800.* Chapel Hill: University of North Carolina Press, 1986.

Lane, Kris E. *Pillaging the Empire.* Armonk, NY: M. E. Sharpe, 1998.

Laurenson, Rae Duncan. "The Emperor Who Smoked a Pipe." *Journal of Laryngology and Otology* 109, no. 1 (January 1995): 1–4.

Le Riverend Brusone, Julio. *La Habana, espacio y vida.* Madrid: Editorial MAPFRE, 1992.

———. *Historia económica de Cuba.* Barcelona: Ediciones Ariel, 1972.

Levine, Robert M. *Cuba in the 1850s: Through the Lens of Charles DeForest Fredricks.* Tampa: University of South Florida Press, 1990.

Lewis, Charles Lee. *Admiral de Grasse and American Independence.* Annapolis: United States Naval Institute, 1945.

Lewis, James A. "Anglo-American Entrepreneurs in Havana: The Background and Significance of the Expulsion of 1784–1785." In Barbier and Kuethe, *North American Role*, 112–26.

Lipson, Leslie. "European Responses to the American Revolution." In "The American Revolution Abroad," special issue, *Annals of the American Academy of Political and Social Science* 428 (November 1976): 22–32.

Lopez Mendez, Lino. *Manual del veguero venezolano: Cultivo del tabaco segun los mejores metodos empleados en Cuba y adaptados a la practica en Venezuela.* Caracas: Imprenta al Vapor de "La Opinion Nacional," 1883.

Lorenzo, Antonio, Marcial Morera, and Gonzalo Ortega. *Diccionario de canarismos.* La Laguna, Tenerife: Francisco Lemus, 1995.

Lorimer, Joyce. "The English Contraband Tobacco Trade in Trinidad and Guiana, 1590–1617." In *The Westward Enterprise: English Activities in Ireland, the Atlantic, and America, 1480–1650*, edited by K. R. Andrews, N. P. Canny, and P. E. H. Hair, 124–50. Liverpool: Liverpool University Press, 1978.

Lucas, George B. *Diseases of Tobacco.* New York: Scarecrow Press, 1958.

Lugar, Catherine. "The Portuguese Tobacco Trade and Tobacco Growers of Bahia in the Late Colonial Period." In *Essays Concerning the Socioeconomic History of Brazil and Portuguese India*, edited by Dauril Alden and Warren Dean, 26–70. Gainesville: University Presses of Florida, 1977.

Luján Muñoz, Jorge. "El establecimiento del Estanco del Tabaco en el Reino de Guatemala." Paper presented at the 5th Congreso Centroamericano de Historia, San Salvador, El Salvador, July 2000.

Luzon, José Luis. "Comer y beber en La Habana colonial." *Caravelle* 71 (December 1998): 23–36.

Lydon, James G. "Fish for Gold: The Massachusetts Fish Trade with Iberia, 1700–1773." *New England Quarterly* 54, no. 4 (December 1981): 539–82.

Lynch, John. *Bourbon Spain, 1700–1808*. Oxford: Blackwell, 1989.

Macías Hernández, Antonio M. *La migración canaria, 1500–1980*. Colombres, Spain: Ediciones Júcar, 1992.

MacInnes, C. M. *The Early English Tobacco Trade*. London: Kegan Paul, Trench, Trubner, and Company, 1926.

Macle Cruz, Jorge. "La tormenta de San Francisco de Borja." *Boletín del Archivo Nacional* 12 (2000): 75.

Madden, Richard R. *La isla de Cuba: Sus recuerdos, progresos y perspectivas*. Translated by Sarah Méndez Capote. 1849. Havana: Consejo Nacional de Cultura, 1964.

Mancall, Peter C. "Tales Tobacco Told in Sixteenth-Century Europe." In *Environmental History* 9, no. 4 (October 2004): 648–78.

Marchena Fernández, Juan. *Oficiales y soldados en el ejército de América*. Seville: Escuela de Estudios Hispanoamericanos, CSIC, 1983.

Marichal, Carlos. *Bankruptcy of Empire: Mexican Silver and the Wars between Spain, Britain, and France, 1760–1810*. Cambridge: Cambridge University Press, 2007.

Marichal, Carlos, and Matilde Souto Mantecón. "Silver and *Situados*: New Spain and the Financing of the Spanish Empire in the Caribbean in the Eighteenth Century." *Hispanic American Historical Review* 74, no. 4 (November 1994): 587–613.

Marrero, Leví. *Cuba: Economía y sociedad*. Vol. 9, *Azucar, ilustración y conciencia (1763–1868) (I)*. Madrid: Editorial Playor, 1983.

———. *Cuba: Economía y sociedad*. Vol. 10, *Azúcar, ilustración y conciencia (1763–1868) (II)*. Madrid: Editorial Playor, 1984.

———. *Cuba: Economía y sociedad*. Vol. 11, *Azúcar, ilustración y conciencia (1763–1868) (III)*. Madrid: Editorial Playor, 1984.

———. *Cuba: Economía y sociedad*. Vol. 12. *Azúcar, ilustración y concencia (1763–1868) (IV)*. Madrid: Editorial Playor, 1985.

———. *Cuba: Economía y sociedad*. Vol. 13, *Azucar, ilustración y conciencia (1763–1868) (V)*. Madrid: Editorial Playor, 1986.

———. *Cuba: Economía y sociedad*. Vol. 7, *Del monopolio hacia la libertad comercial (1701–1763) (II)*. Madrid: Editorial Playor, 1978.

———. *Cuba: Economía y sociedad*. Vol. 8, *Del monopolio hacia la libertad comercial (1701–1763) (III)*. Madrid: Editorial Playor, 1980), 167.

———. *Cuba: Economía y sociedad*. Vol. 4, *El siglo XVII (II)*. Madrid: Editorial Playor, 1975.

Marshall, Woodville K. "Provision-Ground and Plantation Labour in Four Windward Islands: Competition for Resources during Slavery." *Slavery Abolition* 12, no. 1 (May 1991): 48–67.

Martínez-Alier, Verena. *Marriage, Class and Colour in Nineteenth-Century Cuba: A Study of Racial Attitudes and Sexual Values in a Slave Society*. Cambridge: Cambridge University Press, 1974.

Martínez-Fernández, Luis. *Fighting Slavery in the Caribbean: The Life and Times of a British Family in Nineteenth-Century Havana*. Armonk, NY: M. E. Sharpe, 1998.

———. "The 'Male City' of Havana: The Coexisting Logics of Colonialism, Slavery, and Patriarchy in Nineteenth-Century Cuba." In *Women and the Colonial Gaze*, edited by Tamara L. Hunt and Micheline R. Lessard, 104–16. New York: New York University Press, 2002.

Marx, Robert F. *Shipwrecks in the Americas*. 1971. Mineola, NY: Dover, 1987.

Matthee, Rudi. *The Pursuit of Pleasure: Drugs and Stimulants in Iranian History, 1500–1900*. Princeton, NJ: Princeton University Press, 2005.

McCarthy, Charles H. "The Attitude of Spain during the American Revolution." *Catholic Historical Review* 2, no. 1 (April 1916): 47–65.

McFarlane, Anthony. *Colombia before Independence: Economy, Society, and Politics under Bourbon Rule*. Cambridge Latin American Studies. Cambridge: Cambridge University Press, 1993.

McNeill, John Robert. *Atlantic Empires of France and Spain: Louisbourg and Havana, 1700–1763*. Chapel Hill: University of North Carolina Press, 1985.

McWaters, David Lorne. "The Royal Tobacco Monopoly in Bourbon Mexico, 1764–1810." PhD diss., University of Florida, 1979.

Meining, D. W. *The Shaping of America: A Geographical Perspective on 500 Years of History*. Vol. 1, *Atlantic America, 1492–1800*. New Haven, CT: Yale University Press, 1986.

Mendez, Lino Lopez. *Manual del veguero venezolano: Cultivo del tabaco segun los mejores metodos empleados en Cuba y adaptados a la practica en Venezuela*. Caracas: Imprenta al Vapor de "La Opinion Nacional," 1883.

Middleton, Arthur Pierce. "The Chesapeake Convoy System, 1662–1763." *William and Mary Quarterly*, 3rd series, 3, no. 2 (April 1946): 182–207.

Minor, Peter. *Notes on the Cultivation and Management of Tobacco, from the Plant Bed to the Prize; According to the Most Approved Practices in Albemarle, and the Adjacent Counties in Virginia*. Baltimore: J. Robinson, 1822.

Mintz, Sidney W. *Sweetness and Power: The Place of Sugar in Modern History*. New York: Penguin, 1986.

Mohl, Raymond A. ed. "A Scotsman in Cuba, 1811–1812." *Americas* 29, no. 2 (1972): 245.

Mora de Tovar, Gilma Lucía. *Aguardiente y conflictos sociales en la Nueva Granada durante el siglo XVIII*. Bogotá: Universidad Nacional de Colombia, Centro Editorial, 1988.

Moreno Fraginals, Manuel. *El ingenio: Complejo económico social cubano del azúcar*. 1964. Havana: Editorial de Ciencias Sociales, 1978.

———. *The Sugarmill: The Socioeconomic Complex of Sugar in Cuba, 1760–1860*. Translated by Cedric Belfrage. New York: Monthly Review Press, 1976.

Mostert, Noel. *The Line upon a Wind: The Great War at Sea, 1793–1815*. New York: Norton, 2007.

Nash, Robert C. "The English and Scottish Tobacco Trade in the 17th and 18th Centuries: Legal and Illegal Trade." *Economic History Review* 35, no. 3 (1982): 354–72.

Náter, Laura. "Colonial Tobacco: Key Commodity of the Spanish Empire, 1500–1800." In Topik, Marichal, and Frank, *From Silver to Cocaine*, 93–117.

———. "El negocio colonial de tabaco en el Imperio Español." Unpublished manuscript, 2001.

Nieto y Cortadellas, Rafael. *Dignidades nobiliarias en Cuba*. Madrid: Ediciones Cultura Hispanica, 1954.

Nobiatur, J. F. *El veguero practico y cientifico, o sea Cultivo perfeccionado del tabaco con el método de preparar convenientemente la hoja, para sus más ventajosa elaboracíon, todo con arreglo á los adelantados modernos*. Havana: El Telegrafo, 1874.

Norton, Marcy. "The Business of Tobacco in the Spanish Empire, 1590–1636." Working Paper 99-00, International Seminar on the History of the Atlantic World, 1500–1800. Cambridge, MA: Harvard University, 1998.

———. *Sacred Gifts, Profane Pleasures: A History of Tobacco and Chocolate in the Atlantic.* Ithaca, NY: Cornell University Press, 2008.

Ortiz, Altagracia. *Eighteenth-Century Reforms in the Caribbean: Miguel de Muesas, Governor of Puerto Rico, 1769–76.* East Brunswick, NJ: Associated University Presses, 1983.

Ortiz, Fernando. *Cuban Counterpoint: Tobacco and Sugar.* Translated by Harriet de Onís. With an introduction by Bronislaw Malinowski, prologue by Herminio Portell Vilá, and new introduction by Fernando Coronil. 1947. Durham, NC: Duke University Press, 1995.

O'Shaughnessy, Andrew J. "The Formation of a Commercial Lobby: The West India Interest; British Colonial Policy and the American Revolution." *Historical Journal* 40, no. 1 (March 1997): 71–95.

Parcero Torre, Celia María. *La pérdida de la Habana y las reformas borbónicas en Cuba, 1760–1773.* Valladolid, Spain: Consejo de Castilla y León, Consejería de Educación y Cultura, 1998.

Parsons, James J. *Hispanic Lands and Peoples: Selected Writings of James J. Parsons.* Edited by William M. Denevan. Dellplain Latin American Studies. Boulder, CO: Westview Press, 1989.

Percy, David O. "Ax or Plow? Significant Colonial Landscape Alteration Rates in the Maryland and Virginia Tidewater." *Agricultural History* 66, no. 2 (Spring 1992): 66–74.

Perdomo, José E. *Léxico tabacalero cubano.* 1940. Miami: Ediciones Universal, 1998.

Pérez, Louis A., Jr., ed. *Cuba and the United States: Ties of Singular Intimacy.* 3rd ed. Athens: University of Georgia Press, 2003.

———. *Cuba between Empires, 1878–1902.* Pittsburgh: University of Pittsburgh Press, 1998.

———. *Cuba: Between Reform and Revolution.* 4th ed. New York: Oxford University Press, 2010.

———. *Lords of the Mountain: Social Banditry and Peasant Protest in Cuba, 1878–1918.* Pittsburgh: University of Pittsburgh Press, 1989.

———. *On Becoming Cuban: Identity, Nationality, and Culture.* Chapel Hill: University of North Carolina Press, 1999.

———. *Winds of Change: Hurricanes and the Transformation of Nineteenth-Century Cuba.* Chapel Hill: University of North Carolina Press, 2001.

Pérez-Beato, Manuel. *Habana antigua.* Vol. 1, Toponimia. Havana: Seoane, Fernández, 1936.

Pinkerton, John. *A General Collection of the Best and Most Interesting Voyages and Travels, in all Parts of the World; Many of which are now translated into English.* Vol. 5. Philadelphia: Kimber and Conrad, 1812.

Piqueras, José A. "Los amigos de Arango en la Corte de Carlos IV." In González-Ripoll and Álvarez Cuartero, *Francisco Arango*, 151–66.

Porter, Robert Percival. *Appendix to the Report on the Commercial and Industrial Condition of the Island of Cuba.* Washington, DC: Government Printing Office, 1899.

Poyo, Gerald E. "Cuban Patriots in Key West, 1878–1886: Guardians at the Separatist Ideal." *Florida Historical Quarterly* 61, no. 1 (July 1982): 20–36.

Price, Jacob M. *France and the Chesapeake: A History of the French Tobacco Monopoly, 1674–1791, and of Its Relationship to the British and American Tobacco Trades.* 2 vols. Ann Arbor: University of Michigan Press, 1973.

Psoter, Walter, and Douglas E. Morse. "Annual Per Capita Apparent Consumption of Tobacco Products in the United States: 1900–1990." *Preventative Medicine* 32 (2001): 1–9.

Quiroz, Alfonso W. "Cambio en Güines en la era de Arango y Humboldt." In González-Ripoll and Álvarez Cuartero, *Francisco Arango*, 229–44.

———. "Implicit Costs of Empire: Bureaucratic Corruption in Nineteenth-Century Cuba." *Journal of Latin American Studies* 35, no. 3 (August 2003): 473–511.

Ravenholt, R. T. "Tobacco's Global Death March." *Population and Development Review* 16, no. 2 (June 1990): 213–40.

Rediker, Marcus. " 'Under the Banner of King Death': The Social World of Anglo-American Pirates, 1716 to 1726." *William and Mary Quarterly*, 3rd series, 38, no. 2 (April 1981): 203–27.

Rivero Muñiz, José. *Tabaco: Su historia en Cuba.* Vols. 1–2. Havana: Instituto de Historia, 1964–1965.

———. *Las tres sediciones de los vegueros en el siglo XVIII.* Havana: Academia de la Historia de Cuba, 1951.

Rodríguez Gordillo, José M. *Un archivo para la historia del tabaco.* Madrid: Jacaryan, 1984.

———. *La diffusion del Tabaco en España: Diez estudios.* Seville: Universidad de Sevilla, Fundación Altadis, 2002.

Rodríguez Gordillo, José M., José Morales Sanchez, and Alfonso E. Pérez Sánchez. *Goya: Retratos para la Real Fábrica de Tabacos de Sevilla.* Seville: Edita Tabapress para Tabacalera, 1985.

Rodríguez-Ferrer, Miguel. *El tabaco habano: Su historia, su cultivo, sus vicisitudes, sus más afamadas vegas en Cuba.* Madrid: Imprenta del Colegio Nacional de Sordo-Mudos, 1851.

Rodríguez-Salgado, M. J. *The Changing Face of Empire: Charles V, Philip II and Habsburg Authority, 1551–1559.* Cambridge: Cambridge University Press, 2008.

Ronning, C. Neale. *José Martí and the Émigré Colony in Key West: Leadership and State Formation.* New York: Praeger, 1990.

Saether, Steinar A. "Bourbon Absolutism and Marriage Reform in Late Colonial America." *Americas* 59, no. 4 (April 2003): 475–509.

Salinger, Pierre. "Kennedy, Cuba, and Cigars." *Cigar Aficionado* 1, no. 1 (Autumn 1992), *www.cigaraficionado.com.*

———. *P.S.: A Memoir.* New York: Macmillan, 2001.

Saloutos, Theodore. "Efforts at Crop Control in Seventeenth Century America." *Journal of Southern History* 12, no. 1 (February 1946): 45–66.

Salvucci, Linda K. "Anglo-American Merchants and Stratagems for Success in Spanish Imperial Markets, 1783–1807." In Barbier and Kuethe, *North American Role*, 127–33.

Sanders, G. Earl. "Counter-Contraband in Spanish America: Handicaps of the Governors in the Indies." *Americas* 34, no. 1 (July 1977): 59–80.

Santovenia, Emeterio S. *Pinar del Río.* Mexico City: Fondo de Cultura Economica, 1946.

Sanz Rozalén, Vincent. "Los negros del Rey: Tabaco y esclavitud en Cuba a cominenzos del siglo XIX." In *Trabajo libre y coactivo en sociedaded de plantación*, edited by José A. Piqueras, 151–76. Madrid: Siglo XXI Editores, 2009.

Saunt, Claudio. *West of the Revolution: An Uncommon History of 1776*. New York: Norton, 2014.

Schroeder, Susan. *Cuba: A Handbook of Historical Statistics*. Boston: G. K. Hall, 1982.

Schwartz, Rosalie. *Lawless Liberators: Political Banditry and Cuban Independence*. Durham, NC: Duke University Press, 1989.

Schwartz, Stuart B. *Slaves, Peasants, and Rebels: Reconsidering Brazilian Slavery*. Urbana: University of Illinois Press, 1992.

Scott, James C. *Domination and the Arts of Resistance: Hidden Transcripts*. New Haven, CT: Yale University Press, 1990.

Scott, Rebecca J. *Slave Emancipation in Cuba: The Transition to Free Labor, 1860–1899*. Princeton, NJ: Princeton University Press, 1985.

Shafer, Robert J. "Ideas and Work of the Colonial Economic Societies, 1781–1820." In *Revista de Historia de América* 44 (December 1957): 331–68.

Shammas, Carole. *The Pre-Industrial Consumer in England and America*. Oxford: Oxford University Press, 1990.

Silver, Timothy. *A New Face on the Countryside: Indians, Colonists, and Slaves in South Atlantic Forests, 1500–1800*. Cambridge: Cambridge University Press, 1990.

Sintesis de geografia de Cuba. Miami: Ediciones del Directorio Magisterial Cubano (en el exilio), n.d.

Starna, William A., and John H. Relethford. "Deer Densities and Population Dynamics: A Cautionary Note." *American Antiquity* 50, no. 4 (October 1985): 825–32.

Stern, Steve J. *The Secret History of Gender: Women, Men, and Power in Late Colonial Mexico*. Chapel Hill: University of North Carolina Press, 1995.

Stubbs, Jean. "El Habano: The Global Luxury Smoke." Commodities of Empire Working Paper No. 20. Milton Keynes, UK: Open University Ferguson Centre of African and Asian Studies; London: Institute of the Americas, September 2012, *www.open.acu.uk*.

———. *Tabaco en la periferia*. Havana: Editorial de Ciencias Sociales, 1989.

Suárez Argüello, Clara Elena. *Camino real y carrera larga: La arriería en la Nueva España durante el siglo XVIII*. Mexico City: Centro de Investagiciones y Estudios Superiores en Antropología Social, Ediciones de la Casa Chata, 1997.

Sullum, Jacob. *For Your Own Good: The Anti-Smoking Crusade and the Tyranny of Public Health*. New York: Simon and Schuster, 1999.

Tatham, William. *An Historical and Practical Essay on the Culture and Commerce of Tobacco*. Facsimile reprinted in Herndon, *William Tatham*, 1–332.

Thomas, Hugh. *Cuba, or the Pursuit of Freedom*. Updated ed. New York: Da Capo Press, 1998.

Tone, John Lawrence. *War and Genocide in Cuba, 1895–1898*. Chapel Hill: University of North Carolina Press, 2006.

Topik, Steven, Carlos Marichal, and Zephyr Frank, eds. *From Silver to Cocaine: Latin American Commodity Chains and the Building of the World Economy, 1500–2000*. Durham, NC: Duke University Press, 2006.

Topping, Aileen Moore. "Alexander Guillon in Havana, 'This Very Friendly Port.'" *South Carolina Historical Magazine* 83, no. 1 (January 1982): 34–49.

Torres Ramírez, Bibiano. *La Compañía Gaditana de Negros*. Seville: Escuela de los Estudios Hispano-Americanos de Sevilla, 1973.

Tso, T. C. *Physiology and Biochemistry of Tobacco Plants*. Stroudsburg, PA: Dowden, Hutchinson and Ross, 1972.

Tuchman, Barbara W. *The First Salute*. New York: Knopf, 1988.

Tucker, St. George. *Hansford: A Tale of Bacon's Rebellion*. Richmond, VA: George M. West, 1857.

Twinam, Ann. *Public Lives, Private Secrets: Gender, Honor, Sexuality, and Illegitimacy in Colonial Spanish America*. Stanford, CA: Stanford University Press, 1999.

US Customs and Border Protection. "Importing Cuban Cigars." *help.cbp.gov*, accessed 19 June 2014.

US War Department, Office Director Census of Cuba. *Report on the Census of Cuba, 1899*. Washington, DC: Government Printing Office, 1900.

Virginia Department of Emergency Management. "Winter Weather: Virginia Winters." *www.vaemergency.gov/news/history/weather*, accessed 19 October 2012.

Warren, John Esaias. *Vagamundo; or the Attaché in Spain: Including a Brief Excursion into the Empire of Morocco*. New York: Scribner, 1851.

West, Robert C., and John P. Augelli. *Middle America: Its Land and Peoples*. 3rd ed. Englewood Cliffs, NJ: Prentice Hall, 1989.

Whigham, Thomas L. "Agriculture and the Upper Plata: The Tobacco Trade, 1780–1865." *Business History Review* 59, no. 4 (Winter 1985): 563–96.

Whitbeck, R. H. "Geographical Relations in the Development of Cuban Agriculture," *Geographical Review* 12, no. 2 (April 1922): 223–40.

Wilbert, Johannes. *Tobacco and Shamanism in South America*. New Haven, CT: Yale University Press, 1987.

Williams, Neville. *Contraband Cargoes: Seven Centuries of Smuggling*. Camden, CT: Shoe String Press, 1961.

Winter, Joseph C. "Botanical Description of the North American Tobacco Species." In Winter, *Tobacco Use*, 87–127.

———. "From Earth Mother to Snake Woman: The Role of Tobacco in the Evolution of Native American Religious Organization." In Winter, *Tobacco Use*, 265–304.

———, ed. *Tobacco Use by Native Americans: Sacred Smoke and Silent Killer*. Norman: University of Oklahoma Press, 2000.

Wurdemann, John George F. *Notes on Cuba*. Boston: James Munroe, 1844.

Wyman, R. H., comp. *Winds, Currents, and Navigation of the Gulf of Cadiz, the Western Coast of the Spanish Peninsula, and the Strait of Gibraltar, compiled from the Most Reliable Authorities*. Washington, DC: Government Printing Office, 1870.

Ziegler, Vanessa Michelle. "The Revolt of the 'Ever-Faithful Isle': The Ten Years' War in Cuba, 1868–1878." PhD diss., University of California, Santa Barbara, 2007.

Index

Numbers in **bold** refer to maps or tables.